Quantum Genet

Quantum Genetics

Quantum Genetics

V.V. Stcherbic
L.P. Buchatsky
National Taras Schevchenko University of Kyiv
Kyiv
Ukraine

Science Publishers
Enfield (NH) Jersey Plymouth

SCIENCE PUBLISHERS
An Imprint of Edenbridge Ltd., British Isles.
Post Office Box 699
Enfield, New Hampshire 03784
United States of America

Website: *http://www.scipub.net*

sales@scipub.net (marketing department)
editor@scipub.net (editorial department)
info@scipub.net (for all other enquiries)

Library of Congress Cataloging-in-Publication Data

Shcherbik, V.V.
[Kvantovaia genetika English]
Quantum genetics/V.V. Stcherbik, L.P. Buchatsky.
p. cm.
Includes bibliographical references.
ISBN-13: 978-1-57808-508-8(alk. paper)
ISBN-10: 1-57808-508-X(alk. paper)
1. Quantum biochemistry. 2. Genetics. 3. Molecular biology. I. Buchatskii, L.P. II. Title.

QP517.!34S5313 2007
572.8'0151--dc22

2006053051

ISBN 978-1-57808-508-8

Published by Science Publishers, Enfield, NH, USA
An Imprint of Edenbridge Ltd.
Printed in India

Preface to the English Edition

The book was first published in Russian. The authors revised the text for the English edition. The little mistakes and inexactitudes of the Russian edition have been corrected. Table 5.1, 6_{ϕ}-configurations of the genetic element 5-Methylcytosine, which was abridged for the Russian edition, has now been represented completely in the English version.

The non-Abelian calibration field theory is the rapidly developing branch of modern theoretical physics, and this could be the basis of quantum genetics. The introduction of non-Abelian group of internal symmetry for the biological structure (e.g., SU(2) for DNA with the isotopic triplet of information nucleotides **T**, **C**, **U**) gives rise to a phenomenon of asymptotic independence of the complementary DNA pairs' interaction potential from Coulomb forces. It is conditioned by the operation of internal symmetry, which contains the transposition of scattering matrix of interacting physical structure. The Coulomb charge of these structures abrogates and the physical structures couple. Hypochromal effect of DNA is the consequence of pushing of the Coulomb field out to the DNA surface.

The concept of marked point is strongly associated with skewing while transforming of the physical structure within the Poincaré group. In this case, the marked point of terminator with the charge −1 is replacing one of the structure elements. The replaced element is interpreted as an 'observer'. Instead of this element its conformal copy is being introduced. Restoration of the initial structure can be carried out via replacement of the field 'ϕ' by its proper meaning. In other words, the number of elements in the structure increases by +1 and annihilation of the terminator takes place. In many cases, marked point is the part of a physical structure to which the internal symmetry transformation cannot be applied. That is why it can serve as a countable basis generator for the quantitative interpretation of the physical process.

The isotopic invariability of the information nucleotides **T**, **C**, **U** is the consequence of the fact, that only these three nucleotides have one $3, 4_\phi$ – genetic code amino acid configuration:

$$\mathbf{T}: -A + D + H, -P + E + H + \phi.$$
$$\mathbf{U}: -V + E + H, -P + D + H + \phi.$$
$$\mathbf{C}: -V + Q + H, -P + N + H + \phi.$$

Among the other tRNA information nucleotides we should mention inosine (**I**), which has only one 5-genetic code amino acid configuration:

$$\mathbf{I}: -N - I - L + F + R$$

and does not have $3, 4_\phi$-configurations.

The doubled six-dimensional Riemann space forms as a result of SU(2) group stratification in the four-dimensional space and in the conformal field of genetic code SU(4) the group occupies two-dimensional subspace.

In avoidance of tangle the signature of real vector space V (p, q) of dimension $n = p + q$ we designate as p^+q^-, where p is the dimension of the positive definite subspace and q is dimension of negative definite subspace in space V. In quantum genetics any covering function of biological structure has the signature because the comparison of covering function with functions, which belong to separate parts of biological structure, gives an example of incomplete cover . Even rather complex organization of a living cell does not give complete covering functions [of separate units of a cell. In this case the negative part of the signature of covering function] determines that freedom, which is a source of development and transformation of cell structures.

The authors hope that this book will be of interest to geneticists, molecular biologists and physicists-theorists and would appreciate any comments and suggestions.

Kiev, Febuary 2007

VALERY STCHERBIC
LEONID BUCHATSKY

Contents

Introduction

All living organisms contain hereditary material – DNA or RNA – which is necessary for their self-reproduction. Being in a biologically active condition, living organisms exchange matter and energy with the environment by reproducing themselves, evolving and interacting with each other.

Living organisms consist of the same atoms as abiotic environment, but possess a unique ability for self-organization and development; hence, **the genetic information encoded in DNA or RNA determines all of the biological processes in an organism and it can also differentiate bio-structures corresponding to their functions**.

The most important biological structures are DNA, RNA and protein. The term '**genetic information**' is connected directly to the information encoded in nucleotide sequence of DNA, RNA as well as in amino acids of some particular proteins – prions. DNA is the most important bio-molecule of chromosomes, RNA can often be the only genomic material of viruses and prions can easily be determined as *chromosomal ideal*.

Quantum genetics is a science about the essence of biological structures on their molecular level, about the microscopic processes of organization and reorganization of biological structures. It is a synthetic science, which uses the mathematical apparatus of quantum field theory, theory of bundles, theory of groups and finite fields, as well as many other results achieved in the theory of superconductivity, synergetics and quantum chemistry. But **the main foundation of quantum genetics is experimental molecular biology**.

What are the basic principles of quantum genetics?

The most important principles of quantum genetics are that a **genetic code does not have any analogue**; genetic information of DNA is a bundle over the covalent basis **and proteins, DNA form double Riemann space with marked point**.

Let us analyze the most important achievements of the quantum field theory. In the theory of non-Abelian gauge fields which is built on the basis of Lagrange function, the properties of many other fields have also been investigated such as electromagnetic field, Yang-Mills field, scalar field, gravitational field, Higgs field, nuclear field, etc. But they have a very limited use for the description of biological processes. Physicists use geometric forms, which are very distant from those of bio-molecules. In general, physical science operates with virtual functions and variables. On the contrary, quantum genetics needs specific mathematics, the variable of which is specific marked point, which often does not have equivalent representations. For example, authors of the work [1] state that it is not possible to introduce fixed representation basis of vector field in the internal (charge) space of physical particle (even complex one). And this is true indeed. But constituents of molecular internal basis in biology are the real atoms **H, C, O**... and there are no substitutes for them, even though equivalent charge configurations of molecules exist.

The most complex quantum theory in contemporary science is undoubtedly quantum theory of gravitational field, which can be used for the explanation of biological processes. The gravitational field has an integer spin 2, and contains only one marked point (identity of inert and gravitational mass) and it is very weak; hence, it is not suitable for the role of main ordering force in biological processes. But gravitational field has two important features: it cannot be shielded, i.e., to be eliminated by any transformation of coordinates, and the field itself does not depend on structural constants of a physical particle. It is desirable to retain these properties for the fields, which describe interactions of bio-molecules. Genetic code does not depend on any peculiarities of the amino acids structure. This feature suggests that it is possible to introduce a certain field, which is similar to gravitational field by its properties, but can play the role of a general theoretical basis for the interpretation of biological processes. And this field is gravitational field but with a spin of 5/2. We affirm that **conformal field of amino acids is identical to quantum gravitational field**.

The **protonic charge of an atom** is the basic numerical characteristic of amino acid conformal field. Neither lepton nor baryon charge of an atom is important in biology, but the protonic one is. How is it possible to distinguish a protonic charge of the atomic nucleus on the background of the electronic cloud? The answer is simple: gravitational field cannot be shielded.

We could try to compare a conformal field of amino acid to a set of monopoles [2]. For doubled Riemann space, it could be possible to introduce *dyons*. We, however, are not going to do it. The potential of conformal field is determined by doubled charge configurations of the amino acids, which are those of genetic code only in the simplest case and realize the representation of much greater protonic charge than those of amino acids. Doubled charge configurations of amino acids in the bundle field of amino acids **GF**(83) are completely identical and their protonic charges are the same.

The amino acid configurations of index k can be equated with the monopole of charge k. Then these k monopoles are divided by conformal field ϕ, which has a carbon base **C**. But each charge configuration of amino acids has doubled protonic charge Q_p, Q_ζ, which are not equal. This very fact distinguishes charge configurations of amino acids from monopoles. Moreover, charge configurations of amino acids contain a terminator which indicates their disintegration. Nevertheless, it is possible to interpretate DNA strands as complex monopoles.

The fact, that the gravitational field cannot be shielded, can be used for explanation of contraction of the two DNA complementary strands to point. Indeed, in case of separated DNA strands they shield each other. The contraction of strands and their spiralization in one direction is the only DNA conformation, when strand shielding is absent. Indeed, when they do not shield each other, the field $\phi = 0$. Here, we can resort to the analogy with Cooper pairs that concern with the theory of superconduction. But, whereas, in superconductors the magnetic field is pushed out on the superconductor's surface (Meissner effect); in the case of DNA complementary pairing, the **Coulomb field is pushed out** on the DNA surface **by an AT** pair (hypochromic effect). At the same time, a bundle of the field **GF**(137) of inverse fine structure constant occurs and metric $2^+ 2^-$ appears in the DNA internal space. Only two fields **GF**(137) determine the accuracy of **AT** and **GC** pairing. Here, we would like to emphasize that **pushing of the Coulomb field out from the internal space of DNA on its surface by an AT pair is an exceptionally complex natural phenomenon.**

All charge configurations of DNA nucleotides are those of conformal field of amino acids.

But not all amino acids are linearly independent. This situation is common for any Gauge field theory. The condition of amino acid gauge determines 0-vector, the use of which allows calculating an accumulated energy in an integration contour for the Riemann space of the conformal field. The energetic capacity of the contour is equal to the surplus mass of neutrons and the minimal energy resource in the tank contour is determined by a difference of masses between surplus neutrons and a proton.

Actually, the linear dependence between particular amino acids of genetic code freezes their basic charge configurations. We, therefore, introduced a multiple transformation of linear combinations of the amino acids with an orthogonal matrix of transformation – not of separate amino acids, but their subsets. Each subset of amino acids with a symbol 'plus' or 'minus' corresponds to the sequence of amino acids in protein. The rearrangement of amino acids changes the state of protein. But there is no necessity to count on it. The transformation matrix of one junction of the amino acid subsets to another is orthogonal only with regards to the transformation of total protonic charges of these subsets and determines the new subsets of amino acids with a precision up to equivalent charge configurations. That is why the new subsets of amino acids (i.e., proteins) do not have reverse transformation. It means that the gauge transformation matrices of protein structures are integer valued. The most important gauge transformations of protein, which can be presented as a junction of the amino acid subsets, are executed with the use of Hadamard matrices. It assures the growth of protein mass. However, it is also possible to use other integer-valued matrices, for example, in the field **GF**(3). These transformations do not contain introns. The introduction of introns is equivalent to the introduction of marked points into the protein. Transformations of the marked subsets are performed as those of common subsets; but in the final expression for new amino acid subsets, all amino acids with preserved marked points are cut out from protein. The introduction of fractional transformation matrices of protein structures leads to protein disintegration.

Hadamard matrices are the only orthogonal matrices, which transform one protein to another completely. And this is not a fortuity, because Hadamard matrices form Riemann space. Every protein transformation by Hadamard matrix is, nevertheless, irreversible because new protein must have a standard gauge of positions of the amino acid radicals respectively, to its backbone (e.g., antisymmetric) and, therefore, reordering of the amino acids is necessary. But transformation of marked proteins by Hadamard matrix does not require any reordering of amino acids. These transformations are important because they form a number of equifunctional proteins.

Informational nucleotides of DNA and RNA are charge configurations of the genetic code amino acids. An act of RNA translation is, by its essence, a gauge transformation of the conformal field of amino acids. The genetic code transforms one marked protein to another without residue and rearrangement. Hence, Hadamard matrices can be used for building DNA models.

The majority of physicists suppose that the description of physical processes can be performed in space of 1, 2, 3, 4 dimensions. But in biology, it is not only problematic but also impossible to do this. Encoding DNA matrices requires (as a minimum) two more additional degrees of freedom with regards to a four-dimensional space. A time change of the vector of physical system state can lead to the structural differentiation (indexation) of a state's vector, that is, an additional quantization of the multitude of states with formation of real particles and, what is still more important, leads to an additional accumulation of energy in the integration contour of structured vectors of the state. Formation of a mass of Lie group generators of Yang-Mills field is the result of integration. The appearance of mass in generators of vector field is known as Higgs effect. But the phase volume of four-dimensional space decreases with this effect. Non-integrated, marked regions (because they have been already integrated) appear in Hilbert space of vectors (which is full). In order to preserve integration of four-dimensional space, it is necessary to increase the dimension of physical space to six with a metric $4^{+}2^{-}$. Then the DNA coding matrices mark non-integrated zones of four-dimensional space. It is clear that dynamics of such a physical system is very complex.

In the transformation contour of charge configurations of amino acid conform field, basic generators of gauge group perform a role of adapters. There is an analogy between intermediate vector bosons in the Weinberg-Salam theory and pair tRNA-amino acid. But tRNA is a massive particle of amino acid conformal field in a mixed gauge; it transfers the marked point of tRNA-amino acid pairing on the marked point of codon-anticodon pairing in mRNA and, therefore, has a spin of 5/2. The pair tRNA-amino acid is a base-particle having a spin of 2 and is a superconducting system. This conclusion is probably unexpected for biologists and physicists; but the equality of conformal field potentials is possible only in a mixed gauge. Gauge invariance of amino acid conform field breaks in a six-dimensional space but remains in a four-dimensional one.

Using an example of **C**, **N**-trigger of protein, we directly demonstrate the spontaneous breach in symmetry of amino acid conformal field relatively to the time circulation that leads to the acception of information (i.e., new gauge) by an amino acid radical. The analogy of biological processes with physical ones is very conditional.

Adequate description of transformation of genetic information can be performed on the basis of an *ordinal ideal* conception, i.e., an ideal comparison of biological structures. But we also widely use vector representations of conformal field. When describing exchange processes in biology, the charge-exchanging group, isomorphic to Braid group plays an important role.

The charge-exchanging group is determined through invariant charge g by following means:

$$\sum_i G_{iks\ldots p}(x) = \sum_k G_{iks\ldots p}(x) = \ldots = \sum_p G_{iks\ldots p}(x) = g,$$

where sections $\sum_k Giks\ldots p(x) = Gis\ldots p(x_s)\ldots$ realize matrix representation of the group by Latin squares of the order $4g$.

All symmetric groups from S(1) to S(4g) are factor groups of the charge-exchanging group.

Free **FG** group and **YS** group, which form a palindrome on indexes of representations (on ordinal ideals), are root groups of the charge-exchanging group:

$$\mathbf{FG}\,(i,k)s,\ldots,m,n = \mathbf{YS}\,(i,k)n,m,\ldots,s$$

YS group has an obvious representation:

$$(i,k)n,m,\ldots,p,s = (i,k)n,m,\ldots p\,(k,s)n,m,\ldots p\,(s,i)n,m,\ldots p;$$

alternative:

$$k(i,s) = (i,k,s) = (k,s,i) = (s,i,k) = (i,k)s;$$

forms a ring widening at the right:

$$(i,k)n,m,\ldots,p,s = (i,k,s)n,m,\ldots,p = (i,k,s,p,\ldots m,n);$$

acts from the left as a normal devisor:

$$(i,k)n,m,\ldots,p,s,t = k(i,t)n,m,\ldots,p,s = t,k(i,s)n,m,\ldots,p;$$

has a marked point: $(i,i) = (i)$.

Relationship $(i,k)\,(i,k) = 1$ determines cutting out of marked point with indexes i, k. The properties of charge-exchanging group and **YS** group are examined in detail in [3].

The conformal field of amino acids was first obtained in the work of authors [4] as a Gauge field, which transforms vector space of generating operator of protein synthesis into protein envelope of parvoviruses. We show this in Chapter 1 as an initial unmass Gauge field, which becomes mass gauge field of amino acids based on the principle of conformal infinity (reflections of RNA coding nucleotides on numbers of group elements). RNA nucleotides mark cut-out zones of the Gilbert space of generating operator vectors of protein synthesis. The function of tightening this zone into a point (amino acid) is determined by the empiric genetic code.

In Chapter 2, we examine amino acid conformal field in a six-dimensional space with a signature 4^+2^-. Conformal field is built as a vector field, but in a mixed gauge of vectors and it possesses a double property. A marked point* of terminator, which is equivalent to antiproton and is introduced into charge space.

The proof of the statement that amino acid methionine is the first amino acid of conformal field is based on building the **GF**(83) field bundle and representation of ordinal ideals by amino acid radicals.

The conception of entropy is closely connected with neutral charges. The transcendent logarithm function of conformal field states is replaced by rational function of Q_4(Am) amino acid charge by module of a four-dimensional space. Then entropy is determined as a sum of protonic charges of Galois fields with a basis 2. All entropy fields contain marked point. The basic differentiation of amino acids is executed in **GF**(83) Yang-Mills field.

According to the Wedderburn theorem, any finite body is a field. We build a matrix 'α' of amino acid conformity in different gauges of conformal field. But it is not possible to determine field **GF**(83) without an adapter, i.e., covering Hadamard matrix. The adapter determines metric of amino acid conformal field, inserts itself into matrix 'α', and coincides with tRNA. It is necessary to note that there is **no Yang-Mills field without tRNA**.

In order to prove the identity of amino acid conformal field to quantum gravitational field, we obviously build the representation of Poincaré group on states of conformal field. Charges, which are incident to straight-line segment, are introduced. Doubling of amino acid conformal field in a six-dimensional space occurs on connection elements (amino acids). Next, we prove that a curvature tensor of classical gravitational field also possesses a property for doubling. Twenty components of curvature tensor of gravitational field in a four-dimensional field are identified with 20 amino acids of genetic code.

In our arguments, we often use the well-known fact of the absence of finite projective plane of the order six. **Carbon has a protonic charge equal to 6 and is indispensable base of amino acid conformal field**. The Lorenz group is also indispensable. Since the finite projective plane of the order 10 is absent, there is nothing to replace the Poincaré group. Basic transformations of amino acid conformal field cannot be based on Lie groups with other quantity of parameters. But expansions of conformal field are not limited by this requirement.

Chapter 3 examines basic structures of biological systems: DNA, RNA and protein. Each of these structures forms a bundle over their own covariant basis. DNA and RNA contain basis and informational nucleotides, protein contains sequence of amino acid informational radicals and basic amino acid – glutamine. Covariant basis of DNA consists of a marked nucleotide U_{-1}^* (deoxyribose in furanose form, i.e. furanose ring) with the proper marked point $\mathbf{CH_2}$ and unmarked nucleotide C (phosphate strand). DNA informational nucleotides are bound to the point 1′ of the furanose ring.

Proton as a unit of a conversion of fine structure constant 1/137 is a radical of phosphate basis.

Complementary DNA pairs in pairing contact zone form a distribution of conjugated basic nucleotides C and $U(46_\phi)$, which is one of the virtual protein bases.

A joint of an **AT** pair obviously contains a marked point $\mathbf{CH_2}$ of RNA methylation. During the transformation of DNA into RNA, the point $\mathbf{CH_2}$ is changed to the point **O** of the furanose ring and, thus, thymine turns into uracil.

The process of transformation of primary RNA transcript into informational RNA can be described via the concept of *splicing cone.*

According to the *Birkhoff-Hinchin ergodic theorem* [5], the average by the space of realizations of a random value $\xi(t)$ is equal to the temporal average from one realization $\xi(0)$ with a probability 1. Assuming that a random value takes only four values of DNA informational nucleotides – **A**, **T**, **G**, and **C** in the conformal field ϕ with carbon base **C**, we immediately obtain $\xi(0) = \mathbf{A}$. Nucleotide **A** is a countable basis of informational nucleotides; therefore, the length of poly-**A** is proportional to the integration time of correlation errors of the informational RNA with basis RNA correlation, predicted by ordinal type ω_2. Metric tensor, which excises RNA introns, i.e., sets with null measure, forms an informational RNA homeomorphous to the DNA (+) strand; nevertheless, it does not coincide with geometric tensor of informational RNA. This is clear because DNA contains silent genes as well as satellite zones. Therefore, geometric tensor of RNA is also introduced. Enzyme is the only thing that contains DNA metric tensor and is a nucleus of homeomorphism of controlling DNA. Metric and geometric RNA tensors form a complementary DNA sequence.

RNA splicing cone forms coherency integration during the cycle of RNA. Each integration turn constricts phase space and therefore, the enzyme does not contain introns. We can consider that the enzyme is a marked point in transformation process of DNA into RNA in the splicing cone. Informational RNA and protein are the quotient and reminder of splicing process by a module of the introns excised by the metric tensor of RNA.

The Birkhoff-Hinchin theorem is violated for those DNAs, where exons and introns are not mixed well enough. DNA is not transcribed in this case.

The charges of Q_4(Am) amino acids, relatively to the alternating basis of amino acids $\pm Q_p(29)$, form a tangent DNA space containing 24 DNAs in non-ambiguous and 12 DNAs in ambiguous intron gauge.

Chapter 4 examines the quantization modes of the amino acids conformal field. The basic quantization mode is a construction of linear connection at each point of the conformal field. Operational mode is used when expanding basis of conformal field. Charge alternating group is obviously used when forming instanton site – a point of a doubled Riemann space. A concordance of conformal field metric is performed by the induced exchange charges along each of the half-axes of three-dimensional Euclidean space. An introduction of dominant and basic charge configurations of the genetic code amino acids is caused by the fact that the subspaces of doubled Riemann space are not only isomorphic but also non-homeomorphic; because metric tensor of one of them is equal to 4^+2^- and another to 3^+3^-. The amino acid conformal field itself has a metric tensor 13^+7^-. Homeomorphism of the subspace of the doubled Riemann space in a four-dimensional space is induced by an ideal (L, I) of leucine and isoleucine charge configurations coincidence.

Chapter 5 analyses equivalent charge configurations of the basic and informational nucleotides of DNA, RNA and other biological structures.

Minimal charge configurations of **adenine** and **guanine** are the 6ϕ-index configurations of the genetic code amino acids, which are **complementary** to the minimal 3-, 5-index **thymine** configurations and 5-index **cytosine** configurations. The parity and the length of pairing configurations of informational nucleotides determine the DNA structure.

The method of construction of the equivalent charge configurations for biological structures is one of the basic methods for the analysis of biological processes.

Integration of the stochastic process of DNA transcription with the determined process of RNA translation by amino acid conformal field proves that gravitational field with a spin of 5/2 is the main ordering force of biological processes.

Generating Operator of Protein Synthesis

1 Chapter

Quantum genetics—is a quantum science indeed

In theoretical physics, Lagrange function and Hamiltonian of physical system, which are the generating operators of all its properties, play an important role. Many canonic ways have been developed for the energy levels quantization, diagram technique of amplitude calculation of particles scattering as well as many other things for quantum systems. The progress of quantum field theory has been achieved on the basis of the Gauge fields theory. But the use of mathematical apparatus for the description of biological processes is very limited.

The operations of differentiation and integration of biological structures are performed not only on continuous but also on discrete coordinates of their components. Moreover, discrete transformations of the molecules – the biochemical reactions – play a determinative role in the interpretation of biological processes. Therefore, the space of dynamic variables, even of the single bio-molecule, is very complex. Therefore, it makes sense to introduce a generating operator of some biological process instead of the Lagrange function and to synthesize the biological structures 'manually' and not on the basis of 'movement equations.' This is a **combinative-logic** method and in combination with an **analytical** one, it gives the possibility to select the most important aspect of biological processes. And, finally, such a method assumes the use of mathematical structures as constructive elements – representatives of bio-molecules. It means that instead of complete cover space (e.g., Gilbert space) suitable for the description of arbitrary biological process, we use only the finite set of functions, limited volume of mathematical structures, incomplete groups..., i.e., functional space is not static. And this is necessary because bio-molecules perform certain functions during finite time intervals.

Relations, superimposed on dynamic variables of biological structures are of great importance for description of biological processes. In theoretical physics, relations are usually noted as a finite system of equations and play an auxiliary role, whereas in quantum genetics, the scheme of relations between separate structural elements of biological system determines its dynamics. Hence, the description of these relations also assumes dynamic transformation of the biological system.

Quantum genetics needs a mathematical apparatus, which takes into account the most important differences between biological and physical systems. First of all, it is necessary to select appropriate set of mathematical objects for DNA modelling. Four basic coding nucleotides of DNA: adenine **A**, thymine **T**, guanine **G**, and cytosine **C** can be described by any function, which has four states. It would be possible, with the aid of arbitrary generating function by module 4, to generate **DNA code** as a sequence of four numbers 0, 1, 2, 3 comparable with complementary **AT**, **TA**, **CG**, **GC** pairs in one of variants of group S(4), i.e., to select the **DNA gauge**.

The most important interpretation of genetic code among others is the geometric interpretation:

- numbers 0, 1, 2, 3 are determined as indexes of Euclidean coordinates x_0, x_1, x_2, x_3, with discrete values determined from the beginning of DNA double helix;
- DNA is a memory of scale points by four coordinate axes in Euclidean space-time;
- Equal scales by coordinate axes can be obtained only in Riemann space.

However, this method is less universal than that of generation of two sequences from two elements +1 and –1, because these sequences can also be independent. In this case, a record of complementary pair, for example, **A+T+** virtually duplicates Riemann DNA space: unsigned complementary pairs are determined in one of the 24 possible gauges.

In order to obtain uniform scales by all coordinate axes in four-dimensional Euclidean space-time, it is necessary to introduce two functions of DNA transformation:

- **First** function: Four-index determines genetic structure of DNA, RNA splicing and is related to factorization of the first Riemann space;
- **Second** function: Three-index transforms RNA into protein and attaches free first Riemann space to the Riemann space of protein. Protein also forms doubled Riemann space. Thus, we can conclude that protein as well as DNA contains introns.

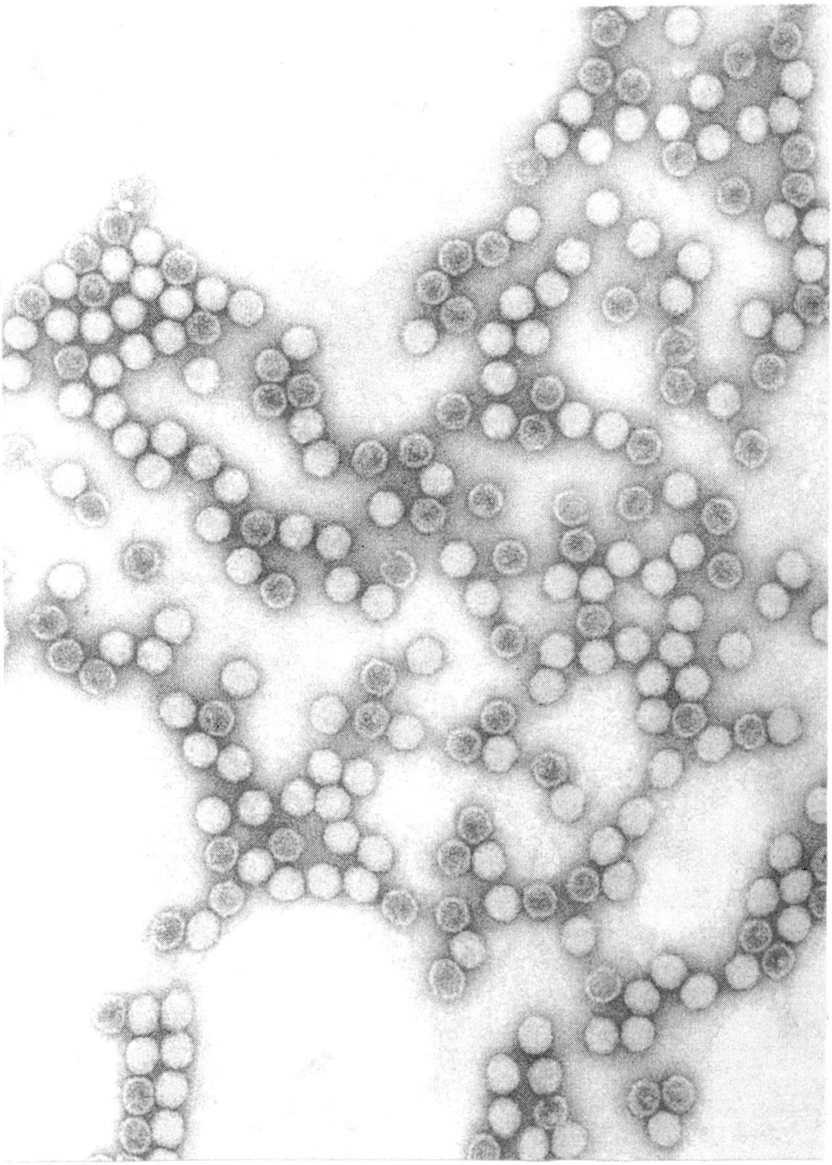

Plate 1 Mosquito densonucleose virus. Diameter of virions 20 nm (Buchatsky, 1982).

It is also necessary to take into account that DNA forms a helical structure and it is advisable that basic set of elements, which is under action of generating function, has a structure of group with torsion. Then, the modelled DNA, with any mode of ordering complementary pairs, will contain information about this group.

As a basic set of elements, we can take Hadamard matrices built of two elements +1 and –1. Hadamard matrices are presentation of Braid group, incident matrices and cell spaces; they are joined matrices of orthogonal transformations and have many uses. Properties of Hadamard matrices are examined in detail in [3]. It is easy to build a Riemann space – a matrix cube (primary instanton) on the basis of Hadamard matrices. Even Penrose twistor transformations have a simple form and give an opportunity to find a holonomy connection group of Hadamard matrix. This fact is determinative argument for the choice of Hadamard matrices as the initial generating set of DNA.

Genome of Aedes DNV virus

In order to build an operator of protein synthesis, which forms a protein on the basis of RNA-encoded genetic information, it is necessary to select an initial DNA with well-known group of spatial symmetry of synthesized structural proteins.

Parvoviruses are the simplest biological objects and are of interest for building biological models of genetic information transformation [6]. Protein envelope of the Aedes DNV (densonucleose virus) has a shape of an icosahedron, the centre of which contains a linear DNA having a length of 4,009 nucleotides (Plate 1)[7].

The model of viral genome transformation into the protein envelope is based on the assumption that genetic information is a presentation of matrix cube of gauge h-field of Hadamard matrices.

Twistor space of h-field transformations includes a transformation of the similar $\pi_\lambda h \pi_\lambda^{-1}$, a proper twistor transformation $\pi_\mu^* h \pi_\mu^{-1}$ of the generator **Ys** of Hadamard matrix bundle. Indexes λ and μ are related to general gradient transformations and transformations by holonomy connection group of matrix h.

The use of matrices π_μ gives an opportunity to turn from matrices h to transposed matrices h^T. Operator **Ys** includes a turn of matrix h on an angle

π around its centre. Moreover, groups **Ys**, **Ys,s** and **Y˜s** can be used for building formulas of the generating genes of DNA. The properties of nucleotides complementarity and DNA strands antisymmetry can be covered by transformations π_μ and **Ys** – it does not matter what term is used to express them.

Pairing of DNA nucleotides is caused by bundle of h-field H136 bound with a fine structure constant by formula α^{-1} = H136 $\subset$ **GF**(137). Each DNA strand induces conjugation because there is a transformation of the initial strand into another, complementary to it; the sum of these strands forms some closed space, the representation of which is an orthogonal matrix cube, which we will call **instanton.**

Each new bundle of a matrix cube can also be described with the help of Hadamard matrices. It is possible to obtain introns of arbitrary complexity using the operator **Ys**. The fact that **DNA doubling in the mode of 'unwinding roll' is also a gene** is a non-trivial result of the use of operator Ys.

Generating operator of protein synthesis includes a ribosome complex, tRNA, set of 20 observed amino acids and basic nucleotides **A**, **U**, **C**, **G**. Supersymmetrical algebra of generating operator has **own central charges – RNA**, but it is induced when reading genome of viral particle. Improper central charges of supersymmetrical algebra are viral DNA. Figure 1.1 shows basic nucleotide sequence of Aedes DNV genome; Table 1.1 contains corresponding statistics.

Let us assume that the length of a linear DNA strand of viral particle is equal to m^3 and is determined by expansion of matrix cube of gauge h-field on the basis of Hadamard matrix of the order m. To generate a genome evaluation of Aedes DNV, let us select a Hadamard matrix and cover the gene in such a way so that the spectrums of coincidences of viral DNA and evaluation DNA nucleotides have a similar structure.

Let us select a covering gene of the test DNA as a relation of two commutators of gauge h-field:

$$\text{covering gene} = \begin{bmatrix} (i,f)\cdot(f,i) \\ (k,f)\cdot(f,k) \end{bmatrix}, \text{ where } (i, f) = h(i, f),$$

in the gauge of nucleotides

$$\begin{matrix} \mathbf{A+T-G-C+} \\ \mathbf{T+A-C+G-}. \end{matrix}$$

```
TATAAGTCCATATTCCATATAAGAAATATTATTTCGTGATACGGATACTGTAAGATACAGTTTCTATTAGAAACGA
TGTATTACATCTGTATCTTACAGTATCCGTATCACGAAATAATATTTCTTATATGGAATATGGACTTATATCAAAG
TTCTATATGGATCACTGGAGGTGGAAAATAAGAGAGAAACATAAGGTGGAAAATAACTTATTATCCACATACAAAT
ACATCCTTAATTTCCACTACCACATGGTCCACCCCTATATAAGGAGTACAAAAGGAGGGCCAAATCGAGTGATGAA
TTCAGTCTGCGTTGAACATTCACCGTGTGAACACGGAAATCTATTTTGTGAGTGCATATATTGTTGGGAGCATGAC
GGTCAGTGCAGGGGGAGAAAATTGGATTTGGGAGCATCAACTGGAATCGAAAGAAGATTGGCCAACGATAACCAAC
AACCAGGGCTCTCAGATTTATATTGCACCGAGACAATACATCTTGCAACTGCAATACCAGAAAGGAGAACCATCGA
TCGAGAAAATTACGTCAAAGATTTCGCTGGTCAAACCGTTGGTGACCTCTACCCACAATTACAAGGCAGCACCGGA
GCCTCTGAACCAATTGATTTCGCATTTCCAACTGTTGGCTCAGGAAGCTGGGAAATACTTGTACGTGAATCTCACA
AACATTTCGAGCCAAATTATACGGAAGAAGCTTATCAATCACATATTAGAAGTGTACGAAGAAGATTATTCCCCGA
AGAAACTATGGATAATAACGGGTCACAGGCAAGCACGACCGAAATGCTACGAGACGCTGTCCAAAGATGCGGTTTT
GAAGGCCCTCCTAACAGCCCAAGCGAAAATAACAGAGATGGAATTGATGGAACGTGTATATCAACCGTGGACATAC
AAAGCAATTGTATTGTTAACGCACATTGCCCAAAACAAGGAACAAGCAATCAAACCAACAAGAGAAAGAAATCAAC
CGATACAACAGAATCAAGCGGATCCAAAAAAAATAAAAGCAGCAATTATCAACAAAATTTACAAGAACAAGGCAGT
ACCAGCATCTCCGACACAATCGATATCGTCGACGGAGAGTTGGATGGATCAACTGGATCGAATCGAGAAACAGCAT
ACTACTCATTCGTCCTCCACAAAAACAACGTTAAAGAGGACTGGAGATACATCGCCACAACCAGGGCCAAGCAAGC
GCCGAGTTTCATCACATTCGATCACGGGGATCACATCCATATCCTCTTCTCCTCGTCAAATACAGGAGGAAACAGC
ACAAGAGTCAGAACCAGAATCACCAAGTTTCTTAGTGCAACAAGCGCAGGAAGTGCAGAAGCGACTATCACATTTT
CCAAAGTTAAATTTCTCAGGAACTACATTCTCTATTGCATCCGTTACGGTATCGAAACAGTCAATATCTATGGAAA
TAAAATCCAACAACAATTAACAGAAGCGATGGATACGTTTAAAATATTATTTGAAAACAGAGACCCGAATGACGTA
ATATTAGAAGCCGGTTGCAAATTATACCACGAAGAAAAAAAAGATAATAAACAAAAAAGATGTGGACAACGGAAAC
AACAAAATCTAACGGACATTATATTGGAAAAAATTAAAGAAAAGAAAATTACAACGGCTCAGCAATGGGAAAACCA
AATAGAACCGGAATTCAAAATACAATTAATGAAAGAGTTTGGATTAAATGTGGACAGTTATGTAACCAGAATAGTA
CGCATCGAAAGAACACGTATACAACAATTGATAAAGGCAAAAACACTTACGGAAATAATGCTTGAAATATTAAATG
ATGAATATATAAAACACTTTACACCAGGAGAAGACAACAGCAAAACAGCAAAATGTATTGAATGGATCGAATATTT
ATTCAAAGAAAATAACATCAATATCATCCACTTCCTGGCATGGAATGAAATTATAAAAACAAAAAGATATAAAAAA
ATAAACGGAATGGTACTAGAAGGGATCACAAACGCAGGAAAATCATTAATATTAGACAACTTATTGGCCATGGTAA
AACCAGAAGAAATACCACGAGAAAGAGACAACAGTGGATTCCACCTTGACCAAGTACCAGGAGCAGGATCAATCCT
ATTTGAAGAACCAATGATAACACCAGTAAACGTCGGAACATGGAAATTATTACTAGAAGGAAAAACCATAAAAACG
GATGTAAAAAACAAAGACAAAGAACCGATAGAACGAACACCAACGTGGATCACAACAGCAACTCCAATAACAAATA
ACATTGATATGAATGAGACATCACAAATACTACAAAGAATAAAACTATATATATTGAAAAAGAGTATCCAACACAG
AGACGACAAATATACTATAAATGCGCAAATACAAAATAAATTAATCAGTCGTCCTCCAACTCTCATTGAGCCAATA
CATATGGCCATAGTGTTTATAAAAAATTTCACAAAAATATATAATCTAATAGCAGAAGAAGACAAAGCACACACAG
TAAACGAGAAGGCAATACAAATCAACAACGAAGTGAAAGAAGAAGCAGAATCATGGCAGACAGCACTTCAATGGAC
CATGACGGAGAACAACGAGGAACAAAACGAAAACGAGACGCAGGCGCTGGAGGATCAGGTGCTGGAATTGGCAAAG
GAACAAGCAACTACGTAAAAGAAGGATATGGACCTAATATGAGCGAAATGGTACCAAGAAACATTATGAATAAAGG
CAACCACACGGTATATCATGTAGTAAAGCAGCAAAAATACTTGGACTTCAACTACGTATCAAATCAAAACCCATAT
ATTATTCCATATCAAACGGCAGGATTCTGGGCATCAATGTGGGACCAAACAGACATCGGATCGAATAACACCATTA
ATATAATGAAAGCACTAAATAACGTATCAGTAGGGGTAACATGGATCAAAGGAGAAATCACGTTCGAAGTATATGC
AGTAACAAGACAACGCTTGCTAACGGGGACAACAAACCAAACTACATGGGACTTTGAAACAAGTCAAAACATGTTC
ATCGCAGATGCACACAGAGAACCAGAAAATTTCAACTTGGCAACAGCAGCAGCAACTGGACCACTTGCACAACAAA
CAACACAAACACTACTATTCAATGCAAACAACGACAGATATACAAAATATGAATTACCACAAAGAAACCAGTATAC
AAGAGAATATGACTTCCAACAACTTACAAATAACTACATGTGGAAACCAACAGACATTAGCGCTGCAGCAAACTTT
AGAAGATTGATCCCAATGGCGGAAGGAGTATATACAACAACAGCGGCAACAACTAAAATGGCAGAATTAACAGAAC
AAAAATCAGTATATGCAGGATCAGGCAAAACAACAGAAGCATCACTATTCAGAAATAGAACATCATATCCTAGAAT
GCATATGGCACAACCACAAGTTCCAGATGAAACCGGATACATGAAATTCAGATACCAAGTACGAATGAGTACAAAA
CTACACCTCGTATTTCATCTATACCCAGATTATAGTACATCAACAAACATAGAATACATGGGGAGACAAGTATTGG
AATTACCAGAAGTAACAGCAACAGGAGGAGTGGTAACATGTATGCCGTATGAAATCAAAACTTAAATATTAAATTC
AACTTGTATCAACTATAACACATATATAATCAATAAAGCATTCAAAAAACATATAAGTCAAATTAATATATATCAC
AATAAAATTCCACCTTAAAACATAAGCTTAATTTCCACCTCCGTATTCCACCTCAGAATATTGGCTTAAAATCCAC
CTCCAATGATACAGTTAGGAAGCTAATATTAGTCCGGGATCCCCGTGTGGCCGATAGGCGAGGATCGAAAGCCCAA
ATTTTGATGACGTCACCTCACACACATACCAAAAGCTTTAGTTTCTAATAGAAACAGCGTATTACGCTTAAAGCTT
TTGGTATGTGTGTGAGGTGACGTCATCAA
```

Fig. 1.1 Nucleotide sequence of **Aedes DNV** genome. A1,645, T844, G716, C776.

The h–field matrices of the numerator and denominator of a covering gene can be different, but computer tests have shown that the best result was in the case of coincidence of matrices. The appropriate matrix h0 is shown in Fig. 1.2. Elements of Hadamard and characteristic matrixes are designated

Table 1.1 Coincidence spectrum of **Aedes DNV** genome nucleotides

Nucleotides	*Number of Nucleotide Coincidences*							
	1	*2*	*3*	*4*	*5*	*6*	*7*	*8*
A	603	241	99	40	10	5	1	2
T	491	123	29	5				
G	410	123	13	4	1			
C	519	109	9	3				

as follows: $(+1) \to 1, (-1) \to -$. An introduction of gauge h-field allows us to determine a holonomy connection group of genome, to build superalgebra generators of generating operators for the viral protein envelope synthesis, and to find irreducible representations of the group of generating operator, which converts subgroup of genome symmetry into the group of envelope symmetry.

The input flow of RNA nucleotides is the congruency nucleus of the covering group of protein synthesis operator. The information about a group of an envelope, which is recorded in RNA, modulates a covering group of generating operator in such a way that it reproduces this group on its proper elements using vector space and free components of connection (amino acids).

Let us build a matrix cube (Fig. 1.3) on sections hs:

$$hs(i, k) = \mathbf{Y}s\mathbf{Y}s\ h0(i, k) = (i, k)s,s = (i, s)(s, k)(i, k)0 = (i, s)\tilde{}k$$

Matrix elements $Q_{ik} = hi\ hk^T$ are superalgebra generators of the gauge transformations of matrix h0, **generators of Lorentz discrete transformation** of RNA strand.

Let us introduce a matrix character Q_{ik} (Fig. 1.2):

$$\chi(Q_{ik}) = \begin{cases} +1, \text{if } Q_{ik} \text{ is symmetric,} \\ -1, \text{if } Q_{ik} \text{ is antisymmetric} \end{cases}$$

The matrix of a direct product $\chi(Q_{ik}) \cdot h0(i, k)$ is the newly gauge invariant h-field.

Let us reduce matrix h0 of the test genome to a normal form H16 by canonic transformation (Fig. 1.4). Holonomy connection group K_π (H16) of matrix H16 contains 20,160 elements, among which we can select one from subgroup **A5**. The group **A5** has a congruency nucleus consisting of 16 elements. Irreducible representations of holonomy connection group of h-field are built on equivalency classes of congruency nucleuses $\pi_\omega = \pi_\mu{}^* \pi_\mu{}^{-1}$ [3].

```
1 1 - - - 1 - 1 - 1 1 1 - - 1 -     1 - - - - - - - 1 - - - - 1 - -
1 - 1 1 - - - - 1 - 1 1 - - 1 1     - 1 1 - - - 1 - - 1 1 - 1 1 - -
- 1 - 1 1 1 1 - - - 1 - - - 1 1     - 1 1 1 1 1 1 - - 1 1 - - 1 - -
1 1 - 1 - - 1 - 1 1 - - - 1 1 -     - - 1 1 - 1 - - - 1 - - - - 1 -
- 1 1 - - 1 - - 1 - 1 - 1 1 1 -     - - 1 - 1 1 - - - - 1 1 1 1 1 1
- - - - - 1 1 - 1 1 - 1 1 - 1 1     - - 1 1 1 1 - 1 - 1 - - 1 1 - 1
- - 1 1 1 1 - - - 1 - 1 - 1 1 -     - 1 1 - - - 1 - - 1 1 1 - - - 1
- - - 1 - - - 1 - 1 1 - 1 1 1 1     - - - - - 1 - 1 - 1 1 1 - - 1 1
- 1 - - 1 - - 1 1 - - 1 - 1 1 1     1 - - - - - - - 1 - - 1 1 - - 1
1 - - - 1 - 1 - - - 1 1 1 1 1 -     - 1 1 1 - 1 1 1 - 1 1 - 1 - 1 1
1 1 1 - 1 - - - - 1 - - 1 - 1 1     - 1 1 - 1 - 1 1 - 1 1 1 1 1 1 -
1 - 1 - - 1 1 1 - - - - - 1 1 1     - - - - 1 - 1 1 1 - 1 1 1 - - -
- - 1 - 1 - 1 1 1 1 1 - - - 1 -     - 1 - - 1 1 - - 1 1 1 1 1 - - 1
1 - - 1 1 1 - 1 1 - - - 1 - 1 -     1 1 1 - 1 1 - - - - 1 - - 1 1 1
- 1 1 1 - - 1 1 - - - 1 1 - 1 -     - - - 1 1 - - 1 - 1 1 - - 1 1 -
1 1 1 1 1 1 1 1 1 1 1 1 1 1 1 1     - - - - 1 1 1 1 1 1 - - 1 1 - 1
```

Fig. 1.2 Matrices h0, $\chi(Q_{ik})$.

```
1 1 1 1 1 1 1 1 1 1 1 1 1 1 1 1     1 - - - 1 - 1 - - - 1 1 1 1 1 -
1 - - - 1 - 1 - - - 1 1 1 1 1 -     - - - - - - - - - - - - - - - -
1 - - 1 1 - 1 1 - 1 - 1 - - - 1     - - - 1 - - - 1 - 1 1 - 1 1 1 1
1 1 1 - 1 - - - - 1 - - 1 - 1 1     1 - - 1 1 1 - 1 1 - - - 1 - 1 -
1 - 1 - - - - 1 1 1 - 1 1 1 - -     - - 1 - 1 - 1 1 1 1 1 - - - 1 -
1 1 - - - - 1 1 1 - 1 - 1 - - 1     1 - 1 1 - 1 1 - - 1 1 - 1 - - -
1 1 1 1 1 - - 1 - - 1 - - 1 - -     1 - - - 1 1 - - 1 1 1 - - 1 - 1
1 1 - 1 - 1 - - - - - 1 1 1 - 1     1 - 1 - - - - 1 1 1 - 1 1 1 - -
1 - - - 1 1 - - 1 1 1 - - 1 - 1     - - - - - 1 1 - 1 1 - 1 1 - 1 1
1 - 1 1 - - - - 1 - 1 1 - - 1 1     - - 1 1 1 - 1 - 1 - - - 1 1 - 1
1 1 - 1 - - 1 - 1 1 - - - 1 1 -     1 - 1 - - 1 1 1 - - - - - 1 1 1
1 - - 1 1 1 - 1 1 - - - 1 - 1 -     - - - 1 - 1 1 1 1 - 1 1 - 1 - -
1 1 1 - 1 1 1 - 1 - - 1 - - - -     1 - - 1 1 - 1 1 - 1 - 1 - - - 1
1 - 1 - - 1 1 1 - - - - - 1 1 1     - - 1 - 1 1 - 1 - - 1 1 1 - - 1
1 - 1 1 - 1 1 - - 1 1 - 1 - - -     - - 1 1 1 1 - - - 1 - 1 - 1 1 -
1 1 - - - 1 - 1 - 1 1 1 - - 1 -     1 - 1 1 - - - - 1 - 1 1 - - 1 1

1 - - 1 1 - 1 1 - 1 - 1 - - - 1     - - - 1 - 1 1 1 1 - 1 1 - 1 - -
- - - 1 - - - 1 - 1 1 - 1 1 1 1     1 - - 1 1 1 - 1 1 - - - 1 - 1 -
- - - - - - - - - - - - - - - -     - 1 1 1 - - 1 1 - - - 1 1 - 1 -
1 - - - 1 1 - - 1 1 1 - - 1 - 1     1 1 1 1 1 1 1 1 1 1 1 1 1 1 1 1
1 1 - - - 1 - 1 - 1 1 1 - - 1 -     1 - 1 1 - 1 1 - - 1 1 - 1 - - -
- 1 - 1 1 - - - 1 1 1 1 1 - - -     1 1 - 1 - 1 - - - - - 1 1 1 - 1
1 - - 1 1 1 - 1 1 - - - 1 - 1 -     - - - 1 - - - 1 - 1 1 - 1 1 1 1
- 1 - - 1 1 1 1 - 1 - - 1 1 - -     - - 1 1 1 1 - - - 1 - 1 - 1 1 -
- - - 1 - 1 1 1 1 - 1 1 - 1 - -     1 - - 1 1 - 1 1 - 1 - 1 - - - 1
1 1 - 1 - 1 - - - - - 1 1 1 - 1     - 1 - 1 1 - - - 1 1 1 1 1 - - -
- 1 - - 1 - - 1 1 - - 1 - 1 1 1     - - 1 1 1 - 1 - 1 - - - 1 1 - 1
- - - - - 1 1 - 1 1 - 1 1 - 1 1     - 1 1 1 - 1 - 1 1 1 - - - - - 1
1 - - - 1 - 1 - - - 1 1 1 1 1 -     1 1 1 1 1 - - 1 - - 1 - - 1 - -
1 1 - - - - 1 1 1 - 1 - 1 - - 1     1 - 1 1 - - - - 1 - 1 1 - - 1 1
1 1 - 1 - - 1 - 1 1 - - - 1 1 -     - 1 - 1 1 1 1 - - - 1 - - - 1 1
- 1 - 1 1 1 1 - - - 1 - - - 1 1     1 1 - 1 - - 1 - 1 1 - - - 1 1 -
```

Fig. 1.3 Matrices h1-h4 of a cube.

```
1 - 1 - - - - 1 1 1 - 1 1 1 - -
1 1 - 1 - 1 - - - - - 1 1 1 - 1
1 1 - - - 1 - 1 - 1 1 1 - - 1 -
1 - 1 1 - 1 1 - - 1 1 - 1 - - -
- - - - - - - - - - - - - - - -
- 1 1 - - - 1 - - 1 1 1 - 1 - 1
1 - 1 - - 1 1 1 - - - - - 1 1 1
- 1 1 1 - 1 - 1 1 1 - - - - - 1
1 1 - 1 - - 1 - 1 1 - - - 1 1 -
- - - 1 - - - 1 - 1 1 - 1 1 1 1
- 1 1 1 - - 1 1 - - - 1 1 - 1 -
1 1 - - - - 1 1 1 - 1 - 1 - - 1
1 - 1 1 - - - - 1 - 1 1 - - 1 1
- - - - - 1 1 - 1 1 - 1 1 - 1 1
- - - 1 - 1 1 1 1 - 1 1 - 1 - -
- 1 1 - - 1 - - 1 - 1 - 1 1 1 -
```

```
- - 1 1 1 1 - - - 1 - 1 - 1 1 -
1 - 1 1 - 1 1 - - 1 1 - 1 - - -
1 - 1 - - 1 1 1 - - - - - 1 1 1
1 1 - 1 - 1 - - - - - 1 1 1 - 1
1 - - 1 1 1 - 1 1 - - - 1 - 1 -
1 1 1 1 1 1 1 1 1 1 1 1 1 1 1 1
1 1 - - - 1 - 1 - 1 1 1 - - 1 -
- - - 1 - 1 1 1 1 - 1 1 - 1 - -
- 1 - - 1 1 1 1 - 1 - - 1 1 - -
1 - - - 1 1 - - 1 1 1 - - 1 - 1
1 1 1 - 1 1 1 - 1 - - 1 - - - -
- 1 - 1 1 1 1 - - - 1 - - - 1 1
- - 1 - 1 1 - 1 - - 1 1 1 - - 1
- 1 1 - - 1 - - 1 - 1 - 1 1 1 -
- 1 1 1 - 1 - 1 1 1 - - - - - 1
- - - - - 1 1 - 1 1 - 1 1 - 1 1
```

```
1 1 1 1 1 - - 1 - - 1 - - 1 - -
1 - - - 1 1 - - 1 1 1 - - 1 - 1
1 - - 1 1 1 - 1 1 - - - 1 - 1 -
- - - 1 - - - 1 - 1 1 - 1 1 1 1
- 1 - 1 1 - - - 1 1 1 1 1 - - -
1 1 - - - 1 - 1 - 1 1 1 - - 1 -
- - - - - - - - - - - - - - - -
- - 1 - 1 1 - 1 - - 1 1 1 - - 1
- 1 1 1 - 1 - 1 1 1 - - - - - 1
- 1 - - 1 - - 1 1 - - 1 - 1 1 1
1 1 - 1 - 1 - - - - - 1 1 1 - 1
- 1 1 - - 1 - - 1 - 1 - 1 1 1 -
1 1 1 - 1 - - - - 1 - - 1 - 1 1
1 - 1 - - - - 1 1 1 - 1 1 1 - -
1 - 1 1 - - - - 1 - 1 1 - - 1 1
- - 1 1 1 1 - - - 1 - 1 - 1 1 -
```

```
- - 1 - 1 - 1 1 1 1 1 - - - 1 -
1 - 1 - - - - 1 1 1 - 1 1 1 - -
- 1 - - 1 1 1 1 - 1 - - 1 1 - -
1 1 - - - - 1 1 1 - 1 - 1 - - 1
- 1 1 1 - 1 - 1 1 1 - - - - - 1
- - - 1 - 1 1 1 1 - 1 1 - 1 - -
- - 1 - 1 1 - 1 - - 1 1 1 - - 1
1 1 1 1 1 1 1 1 1 1 1 1 1 1 1 1
1 - 1 - - 1 1 1 - - - - - 1 1 1
1 - - 1 1 - 1 1 - 1 - 1 - - - 1
1 1 1 1 1 - - 1 - - 1 - - 1 - -
- 1 - - 1 - - 1 1 - - 1 - 1 1 1
1 1 - - - 1 - 1 - 1 1 1 - - 1 -
- 1 1 1 - - 1 1 - - - 1 1 - 1 -
1 - - 1 1 1 - 1 1 - - - 1 - 1 -
- - - 1 - - - 1 - 1 1 - 1 1 1 1
```

```
1 - - - 1 1 - - 1 1 1 - - 1 - 1
- - - - - 1 1 - 1 1 - 1 1 - 1 1
- - - 1 - 1 1 1 1 - 1 1 - 1 - -
- 1 1 - - 1 - - 1 - 1 - 1 1 1 -
1 1 - 1 - - 1 - 1 1 - - - 1 1 -
1 - 1 1 - - - - 1 - 1 1 - - 1 1
- 1 1 1 - 1 - 1 1 1 - - - - - 1
- 1 - 1 1 - - - 1 1 1 1 1 - - -
1 1 1 1 1 1 1 1 1 1 1 1 1 1 1 1
1 1 - - - - 1 1 1 - 1 - 1 - - 1
1 - 1 - - - - 1 1 1 - 1 1 1 - -
1 1 1 - 1 1 1 - 1 - - 1 - - - -
1 - - 1 1 1 - 1 1 - - - 1 - 1 -
- - 1 - 1 - 1 1 1 1 1 - - - 1 -
- - 1 1 1 - 1 - 1 - - - 1 1 - 1
- 1 - - 1 - - 1 1 - - 1 - 1 1 1
```

```
1 - 1 1 - - - - 1 - 1 1 - - 1 1
- - 1 1 1 - 1 - 1 - - - 1 1 - 1
1 1 - 1 - 1 - - - - - 1 1 1 - 1
1 - 1 - - 1 1 1 - - - - - 1 1 1
1 1 1 - 1 1 1 - 1 - - 1 - - - -
- 1 1 1 - - 1 1 - - - 1 1 - 1 -
- 1 - - 1 - - 1 1 - - 1 - 1 1 1
- 1 1 - - 1 - - 1 - 1 - 1 1 1 -
1 1 - - - - 1 1 1 - 1 - 1 - - 1
- - - - - - - - - - - - - - - -
1 - - 1 1 1 - 1 1 - - - 1 - 1 -
- - 1 - 1 1 - 1 - - 1 1 1 - - 1
- 1 - 1 1 1 1 - - - 1 - - - 1 1
- - - 1 - 1 1 1 1 - 1 1 - 1 - -
1 1 1 1 1 - - 1 - - 1 - - 1 - -
1 - - - 1 - 1 - - - 1 1 1 1 1 -
```

Fig. 1.3 (cont.) Matrices h5-h12 of a cube.

```
1 1 - 1 - - 1 - 1 1 - - - 1 1 -
1 - 1 - - 1 1 1 - - - - - 1 1 1
- 1 - - 1 - - 1 1 - - 1 - 1 1 1
- - 1 1 1 - 1 - 1 - - - 1 1 - 1
- 1 1 1 - - 1 1 - - - 1 1 - 1 -
1 1 1 - 1 1 1 - 1 - - 1 - - - -
1 1 - 1 - 1 - - - - - 1 1 1 - 1
- - - - - 1 1 - 1 1 - 1 1 - 1 1
1 - 1 - - - - 1 1 1 - 1 1 1 - -
1 - - 1 1 1 - 1 1 - - - 1 - 1 -
- - - - - - - - - - - - - - - -
- 1 - - 1 1 1 1 - 1 - - 1 1 - -
- - 1 1 1 1 - - - 1 - 1 - 1 1 -
- 1 1 1 - 1 - 1 1 1 - - - - - 1
1 - - 1 1 - 1 1 - 1 - 1 - - - 1
1 1 1 - 1 - - - - 1 - - 1 - 1 1
```

```
1 - - 1 1 1 - 1 1 - - - 1 - 1 -
1 1 1 - 1 - - - - 1 - - 1 - 1 1
1 1 1 1 1 - - 1 - - 1 - - 1 - -
- 1 1 1 - 1 - 1 1 1 - - - - - 1
1 1 - - - - 1 1 1 - 1 - 1 - - 1
- 1 - 1 1 1 1 - - - 1 - - - 1 1
- 1 1 - - 1 - - 1 - 1 - 1 1 1 -
1 - 1 1 - 1 1 - - 1 1 - 1 - - -
- - - 1 - - - 1 - 1 1 - 1 1 1 1
1 1 - 1 - - 1 - 1 1 - - - 1 1 -
- 1 - - 1 1 1 1 - 1 - - 1 1 - -
- - - - - - - - - - - - - - - -
1 - - - 1 1 - - 1 1 1 - - 1 - 1
- - 1 1 1 - 1 - 1 - - - 1 1 - 1
- - 1 - 1 - 1 1 1 1 1 - - - 1 -
1 - 1 - - 1 1 1 - - - - - 1 1 1
```

```
1 1 1 - 1 1 1 - 1 - - 1 - - - -
1 - - 1 1 - 1 1 - 1 - 1 - - - 1
- 1 1 1 - 1 - 1 1 1 - - - - - 1
1 1 1 1 1 - - 1 - - 1 - - 1 - -
1 - 1 1 - - - - 1 - 1 1 - - 1 1
1 1 - 1 - - 1 - 1 1 - - - 1 1 -
- - - 1 - 1 1 1 1 - 1 1 - 1 - -
1 1 - - - 1 - 1 - 1 1 1 - - 1 -
- 1 1 - - - 1 - - 1 1 1 - 1 - 1
- 1 - 1 1 1 1 - - - 1 - - - 1 1
- - 1 1 1 1 - - - 1 - 1 - 1 1 -
1 - - - 1 1 - - 1 1 1 - - 1 - 1
- - - - - - - - - - - - - - - -
- 1 - - 1 - - 1 1 - - 1 - 1 1 1
1 - 1 - - 1 1 1 - - - - - 1 1 1
- - 1 - 1 - 1 1 1 1 1 - - - 1 -
```

```
- 1 - 1 1 - - - 1 1 1 1 1 - - -
- - 1 - 1 1 - 1 - - 1 1 1 - - 1
1 1 - - - - 1 1 1 - 1 - 1 - - 1
1 - 1 1 - - - - 1 - 1 1 - - 1 1
- - - - - 1 1 - 1 1 - 1 1 - 1 1
1 - - 1 1 - 1 1 - 1 - 1 - - - 1
- 1 - 1 1 1 1 - - - 1 - - - 1 1
- 1 1 1 - - 1 1 - - - 1 1 - 1 -
- - 1 - 1 - 1 1 1 1 1 - - - 1 -
1 1 1 - 1 - - - - 1 - - 1 - 1 1
- 1 1 1 - 1 - 1 1 1 - - - - - 1
1 1 - - - 1 - 1 - 1 1 1 - - 1 -
1 - 1 1 - 1 1 - - 1 1 - 1 - - -
- - - - - - - - - - - - - - - -
1 1 1 - 1 1 1 - 1 - - 1 - - - -
1 - - 1 1 1 - 1 1 - - - 1 - 1 -
```

```
- 1 - - 1 - - 1 1 - - 1 - 1 1 1
- - 1 1 1 1 - - - 1 - 1 - 1 1 -
1 1 - 1 - - 1 - 1 1 - - - 1 1 -
- 1 - 1 1 1 1 - - - 1 - - - 1 1
1 1 1 - 1 - - - - 1 - - 1 - 1 1
1 - - - 1 - 1 - - - 1 1 1 1 1 -
1 - 1 1 - - - - 1 - 1 1 - - 1 1
1 - - 1 1 1 - 1 1 - - - 1 - 1 -
1 1 - - - 1 - 1 - 1 1 1 - - 1 -
- - - - - 1 1 - 1 1 - 1 1 - 1 1
- 1 1 - - 1 - - 1 - 1 - 1 1 1 -
- - 1 - 1 - 1 1 1 1 1 - - - 1 -
1 - 1 - - 1 1 1 - - - - - 1 1 1
- - - 1 - - - 1 - 1 1 - 1 1 1 1
1 1 1 1 1 1 1 1 1 1 1 1 1 1 1 1
- 1 1 1 - - 1 1 - - - 1 1 - 1 -
```

```
- - 1 1 1 - 1 - 1 - - - 1 1 - 1
1 - 1 1 - - - - 1 - 1 1 - - 1 1
- 1 - 1 1 1 1 - - - 1 - - - 1 1
- - 1 - 1 1 - 1 - - 1 1 1 - - 1
1 - - 1 1 - 1 1 - 1 - 1 - - - 1
- - - - - 1 1 - 1 1 - 1 1 - 1 1
1 1 - - - - 1 1 1 - 1 - 1 - - 1
- - - 1 - - - 1 - 1 1 - 1 1 1 1
- 1 - - 1 - - 1 1 - - 1 - 1 1 1
- 1 1 1 - 1 - 1 1 1 - - - - - 1
1 1 1 - 1 - - - - 1 - - 1 - 1 1
1 - 1 - - 1 1 1 - - - - - 1 1 1
1 1 - 1 - 1 - - - - - 1 1 1 - 1
- 1 1 - - - 1 - - 1 1 1 - 1 - 1
1 - - - 1 1 - - 1 1 1 - - 1 - 1
1 1 1 1 1 1 1 1 1 1 1 1 1 1 1 1
```

Fig. 1.3 (cont.) Matrices h13-h16 of a cube.

```
1 1 1 1 1 1 1 1 1 1 1 1 1 1 1 1    1 1 1 1 1 1 1 1 1 1 1 1 1 1 1 1
1 1 1 - - - - 1 - 1 - - 1 1 - 1    1 1 1 1 1 1 1 - - - - - - - - 1
1 1 - - - - 1 - 1 - - 1 1 - 1 1    1 1 1 - - - - 1 1 1 1 - - - - 1
1 - - - - 1 - 1 - - 1 1 - 1 1 1    1 1 - - - 1 1 1 - - 1 1 1 - - -
1 - - - 1 - 1 - - 1 1 - 1 1 1 -    1 1 - - 1 1 - - 1 1 - - 1 1 - -
1 - - 1 - 1 - - 1 1 - 1 1 1 - -    1 1 - 1 1 - - 1 - - 1 - - 1 1 -
1 - 1 - 1 - - 1 1 - 1 1 1 - - -    1 1 - 1 - - 1 - 1 1 - 1 - - 1 -
1 1 - 1 - - 1 1 - 1 1 1 - - - -    1 - 1 1 - 1 - 1 - 1 - - 1 - 1 -
1 - 1 - - 1 1 - 1 1 1 - - - - 1    1 - 1 - 1 - 1 - 1 - 1 - 1 - 1 -
1 1 - - 1 1 - 1 1 1 - - - - 1 -    1 - 1 - 1 - 1 1 - 1 - 1 - 1 - -
1 - - 1 1 - 1 1 1 - - - - 1 - 1    1 - 1 1 - 1 - - 1 - 1 1 - 1 - -
1 - 1 1 - 1 1 1 - - - - 1 - 1 -    1 - - 1 - - 1 - - 1 1 - 1 1 - 1
1 1 1 - 1 1 1 - - - - 1 - 1 - -    1 - - 1 1 - - 1 1 - - 1 1 - - 1
1 1 - 1 1 1 - - - - 1 - 1 - - 1    1 - - - 1 1 - - - 1 1 1 - - 1 1
1 - 1 1 1 - - - - 1 - 1 - - 1 1    1 - - - - 1 1 1 1 - - - - 1 1 1
1 1 1 1 - - - - 1 - 1 - - 1 1 -    1 1 1 - - - - - - - - 1 1 1 1 1
```

Fig. 1.4 Matrices H16 of **A**5 group and **G**96 group.

Figure 1.5 shows the evaluation nucleotide sequence of **Aedes DNV** genome and Table 1.2 contains the corresponding statistics.

Table 1.2 Coincidence spectrum of the nucleotides of **Aedes DNV** evaluation genome

Nucleotides	*Number of Nucleotide Coincidences*						
	1	*2*	*3*	*4*	*5*	*6*	*7*
A	544	230	123	52	11	6	4
T	398	78	18	1			
G	504	134	36	3			
C	503	130	39	3			

Irreducible representation of icosahedron group A5, group of generating operator of protein synthesis G96

Subgroup A5 of group K_π (H16)

1. (1)(2)(3)(4)(5)(6)(7)(8)(9)(10)(11)(12)(13)(14)(15)(16)
2. (1)(2, 7, 12)(3, 10, 16)(4, 13, 6)(5, 8, 9)(11, 15, 14)
3. (1)(2, 9)(3, 4)(5, 13)(6, 12)(7, 10)(8, 16)(11)(14)(15)
4. (1)(2, 5, 6)(3, 13, 8)(4, 10, 12)(7, 16, 9)(11, 15, 14)
5. (1)(2, 6, 5)(3, 8, 13)(4, 12, 10)(7, 9, 16)(11, 14, 15)
6. (1)(2, 12, 7)(3, 16, 10)(4, 6, 13)(5, 9, 8)(11, 14, 15)
7. (1)(2, 10, 8)(3, 7, 6)(4, 5, 16)(9, 13, 12)(11, 15, 14)
8. (1)(2, 8, 10)(3, 6, 7)(4, 16, 5)(9, 12, 13)(11, 14, 15)

AGAGCACATGGAAACCTGAGACAAGGGACAAAGGAGCACAGGGACCAAGTATAAAATTTAAAAATGCGAACCGGTA
AAACTTATACAAGTTAACCCGGAGACAAGGGAACCATTAGCAAAGTTCCAAAGGAGCACATGGACCACGGAGCCCA
TGGACACCTGAGAAAATGGAAAACGGCTAAAATTTAAACAGGATAACCGTGAAACATTAGAACCGGTACCACTTAG
ACCAGTTCAACATTAGCCACGGTAACCCGAGACATGCAGAAACTTAGAACGGAAGACAAGGAGACATGAAGACCAG
GCGCAAGGCCTAAAAGTATAAATTAATAAAATTCGCACGGACTAACCTGAGAACGGAAGAACCGTCGACAGGACTC
CAAGGAGACATGCAGACCATGAGACCTGCAGACACTTAGAAAGGCAGAAAATGATCAAGGCCTAAACGGAGCAATT
ACGAAACGTCGAAATTAATACCATTCGAACTGACTCAACGTCGACCGTAATAACCTGGAGCGCACGGAAGTTCGAG
ATAAAGGACGGGAGAGCGCAAGGACTGGATATAGAACTTAAGGGCGCGAGCCAGTAAGGTCTATATAAATTAATTT
AGAGATAAAGGAATTGCTAGCGAAATTCCGGGAGAGCGCACGGACTGTAGAGCTCACGGACGTTCGAGAGAACGGA
AGGTAGCTAGAACTTAAGTGAGATAGCCATGAAGTGCTAGAGCCAGTACTGTCTAGATCAATTCAGTGCTAGCTAC
AGTAATTTAAAACGCACAAAAGTCTAAAATAAAAAACGGAGAAACGCAAAAACTGAGCACAGAACCCAAGGATACA
AGCCAACAAGGCTCACATAAACCAATTCGAAAATAAAAAAATTATCAACGAAACCCCGGAGAAACGCACAAACTGC
GAAACTCACAAACGTCTAAAAGAACAAAAGGCGACCAGAACCCAAGTAGAACAGCCACAAAGTATCAAAGCCAACA
CTGCTCAAATCAACCCAGTATCAACTACAACAATTCGGAATACACGGGGACCCGAAGCAAAGGGTAAAAGAATACA
AGGGTCAAATACGAAACTTGGAAACGCAGACCAGTGGAACCTACGCAAATTGGCCCAGAAGCAAAGGGGCCACTAA
TAAAATTTTAAAAGAATACACGGGTCACAGAATCCACGGGTACCCGAAGAAACGGGGAACAGCCGAAACTTGGACA
AGACGACCATGGGACACTAAGACCAGTGTCACCTAAGCCAATTTGACACTAATCACAGTGGCCCAAAATATATAAA
GGCTCAAAGCGAGAAATGAGAAAATATAGAAATTAGACACGAGATCCAGGAGCACAGATCGACAGGATCCACGCGA
GCCAGTCTAAAAGCGAGAAAGTCGCCAATAGAGCCCTGAGAAAATATATAAATTATAAAATCTATAAATGCTCAAA
GAGATAAAGGATAACCGAGATCCAGGCGAAACGATCGCAAGGCGCCAAGATCGACATTATCCAAGCTAGCCCGGCG
CCAATCGCGACAGTCTAAAATGTATAAAGATTTAAAGTGAGAAATAGGGAAATGTAGAAATCGGGCACGGGATCCA
GAGGTACAGGTCGACAGAGTTCACGTGAGCCAGCTTGAAAGTGAGAAAGCTGTCAATGGAGCCCTAGGGAAATGTA
TAAATCGTGAAATTTATAAATATTTAAAGGGATAAAGAGTGACCGGGATCCAGATGGAACGGTCGCAAGATGTCAA
GGTCGACATCGTTCAAGTTAGCCCGATGTCAATTGCGACAGCTTGAAACACATAAAAACTCAAAACAAGAAACAAG
AAAACACAGAAACCAGACACAAAATCCAAAAGCACAAACCGACAAAATCCACACAAGCCAACCTAAAAACAAGAAA
ACCGCCAACAAAGCCCCAAGAAAACACATAAACCATAAAACCCATAAACACTCAAAAAAATAAAAAATAACCAAAA
TCCAAACGAAACAACCGCAAAACGCCAAAACCGACACCATCCAAACCAGCCCAACGCCAACCACGACAACCTAAGG
CACATGGAAATCCAGGACAAGGGACAGAAAGGCACAGGGACCGAACGTAAAATTTAAAGACATGAACCGGTAAAGC
CCGTACAAGTTAACTCAAGGACAAGGGAACTACCGGCAAAGTTCCAGAAAGGCACATGGACCGCAAGGCCCATGGA
CATCCAGGAAAATGGAAAGCAATTAAAATTTAAATAAAGTAACCGTGAAATACCGGAACCGGTACCGCCCGGACCA
GTTCAATACCGGCCACGGTAACTCAAAGCATGCGAAAATCTAAGACGGAGAACAGAGAAGCATGAGAACCGAGCAT
AAGGCTCAAAGATACGAATTAGCAAAGCTCATACGGATCAACTCGAAGACGGAGAAACTATCAGCAGGATCCCAGA
GAAGCATGCGAACCGCGAAGCCTGCGAACATCTAAGAAGGCGAAAAGCGACTAAGGCTCAAATAGAATAATTATAA
AATATCAGAATTAGCACCGCTCAGACTGATCCAATATCAGCCGTAGCAACTCGGAACATGCAGAGGCTTGAAACGG
AAGATGAGGGAACATGAAGATTAGGTACAAGGCCTAGGAGTGCAAATTAATAGGATTTACACGGACTAGTCTGGAA
ACGGAAGAGTCGTTAACAGGACTCTGAGGGAACATGCAGATTATGGAACCTGCAGATGCTTGAAAAGGCAGAGGAT
GGCCAAGGCCTAGGCGGACAATTACGAGGCGTTAAAATTAATATTATTTAAACTGACTCGGCGTTAACCGTAATA
GTCTGGAACGTACGGGAATCTGAAATGAAGGGCAGAGGAACGTAAGGGCCGAGTACAGGACTTGAAGATGCAAGTC
AGTGAAGCTTACATGAATTGACTCGGAAATGAAGGGACTATTAACGGAATTTCAGAGGAACGTACGGGCCGCGGAA
CTTACGGGCATCTGAAAGGACGGGAAGCGGCCAGGACTTGAATAGGACAGTCATGGAATATTAAAGTCAGTGCCGC
TTAAATTAATTTAATATTAACTGCAGTGACTCGGAATGCGCAGAAGTTCGAAGTAGAAGACGGGAGAATGCGAAGA
CTGGATACGGAGCCTAAGGGCGCAGGCTAATAAGGTCTACGTAGACTAATTTAGAAGTAGAAGAATTGCTAATGAG
ACTCCGGGAGAATGCGCAGACTGTAGAATTCGCAGACGTTCGAAGGAGCAGAAGGTAGCCGGAGCCTAAGTGAGAC
GGCTACGAAGTGCTAAGGCTAATACTGTCTAAGTCGACTCAGTGCTAATTATAATAATTTA

Fig. 1.5 Evaluation nucleotide sequence of **Aedes DNV** genome. A1,700, T612, G892, C892.

9. (1)(2, 13, 16)(3, 5, 12)(4, 7, 8)(6, 9, 10)(11, 15, 14)
10. (1)(2, 16, 13)(3, 12, 5)(4, 8, 7)(6, 10, 9)(11, 14, 15)
11. (1)(2, 4)(3, 9)(5, 7)(6, 8)(10, 13)(11)(12, 16)(14)(15)
12. (1)(2, 3)(4, 9)(5, 10)(6, 16)(7, 13)(8, 12)(11)(14)(15)
13. (1)(2, 14, 4, 8, 6)(3, 9, 7, 15, 5)(10, 12, 11, 16, 13)
14. (1)(2, 4, 6, 14, 8)(3, 7, 5, 9, 15)(10, 11, 13, 12, 16)
15. (1)(2, 6, 8, 4, 14)(3, 5, 15, 7, 9)(10, 13, 16, 11, 12)
16. (1)(2, 8, 14, 6, 4)(3, 15, 9, 5, 7)(10, 16, 12, 13, 11)
17. (1)(2, 12, 7)(3, 15, 4)(5, 13, 11)(6, 9, 10)(8, 14, 16)

18. (1)(2, 7, 12)(3, 4, 15)(5, 11, 13)(6, 10, 9)(8, 16, 14)
19. (1)(2, 9, 8, 11, 16)(3, 14, 4, 7, 10)(5, 12, 6, 13, 15)
20. (1)(2, 8, 16, 9, 11)(3, 4, 10, 14, 7)(5, 6, 15, 12, 13)
21. (1)(2, 16, 11, 8, 9)(3, 10, 7, 4, 14)(5, 15, 13, 6, 12)
22. (1)(2, 11, 9, 16, 8)(3, 7, 14, 10, 4)(5, 13, 12, 15, 6)
23. (1)(2, 13)(3, 12)(4)(5)(6, 11)(7, 8)(9, 14)(10, 15)(16)
24. (1)(2, 15, 3, 6, 16)(4, 7, 11, 13, 9)(5, 12, 14, 8, 10)
25. (1)(2, 3, 16, 15, 6)(4, 11, 9, 7, 13)(5, 14, 10, 12, 8)
26. (1)(2, 16, 6, 3, 15)(4, 9, 13, 11, 7)(5, 10, 8, 14, 12)
27. (1)(2, 6, 15, 16, 3)(4, 13, 7, 9, 11)(5, 8, 12, 10, 14)
28. (1)(2, 11, 4)(3, 13, 8)(5, 7, 14)(6, 10, 9)(12, 15, 16)
29. (1)(2, 4, 11)(3, 8, 13)(5, 14, 7)(6, 9, 10)(12, 16, 15)
30. (1)(2, 11, 3, 5, 10)(4, 9, 12, 15, 8)(6, 7, 14, 13, 16)
31. (1)(2, 3, 10, 11, 5)(4, 12, 8, 9, 15)(6, 14, 16, 7, 13)
32. (1)(2, 10, 5, 3, 11)(4, 8, 15, 12, 9)(6, 16, 13, 14, 7)
33. (1)(2, 5, 11, 10, 3)(4, 15, 9, 8, 12)(6, 13, 7, 16, 14)
34. (1)(2, 14, 9, 13, 5)(3, 4, 12, 11, 6)(7, 15, 10, 8, 16)
35. (1)(2, 9, 5, 14, 13)(3, 12, 6, 4, 11)(7, 10, 16, 15, 8)
36. (1)(2, 5, 13, 9, 14)(3, 6, 11, 12, 4)(7, 16, 8, 10, 15)
37. (1)(2, 13, 14, 5, 9)(3, 11, 4, 6, 12)(7, 8, 15, 16, 10)
38. (1)(2, 14, 3)(4, 16, 5)(6, 9, 10)(7, 15, 13)(8, 12, 11)
39. (1)(2, 3, 14)(4, 5, 16)(6, 10, 9)(7, 13, 15)(8, 11, 12)
40. (1)(2, 12, 7)(3, 13, 8)(4, 14, 9)(5, 15, 10)(6, 16, 11)
41. (1)(2, 7, 12)(3, 8, 13)(4, 9, 14)(5, 10, 15)(6, 11, 16)
42. (1)(2, 15, 4, 10, 13)(3, 12, 14, 16, 9)(5, 6, 8, 7, 11)
43. (1)(2, 4, 13, 15, 10)(3, 14, 9, 12, 16)(5, 8, 11, 6, 7)
44. (1)(2, 13, 10, 4, 15)(3, 9, 16, 14, 12)(5, 11, 7, 8, 6)
45. (1)(2, 10, 15, 13, 4)(3, 16, 12, 9, 14)(5, 7, 6, 11, 8)
46. (1)(2, 5)(3, 11)(4, 12)(6)(7, 16)(8, 15)(9)(10)(13, 14)
47. (1)(2, 14)(3)(4, 6)(5, 9)(7, 15)(8)(10, 16)(11, 12)(13)
48. (1)(2, 12, 7)(3, 9, 11)(4, 5, 16)(6, 15, 8)(10, 14, 13)
49. (1)(2, 7, 12)(3, 11, 9)(4, 16, 5)(6, 8, 15)(10, 13, 14)
50. (1)(2, 16)(3)(4, 7)(5, 12)(6, 15)(8)(9, 11)(10, 14)(13)
51. (1)(2, 10)(3)(4, 15)(5, 11)(6, 7)(8)(9, 12)(13)(14, 16)
52. (1)(2, 15)(3, 16)(4, 13)(5, 8)(6)(7, 11)(9)(10)(12, 14)
53. (1)(2, 8)(3, 7)(4, 14)(5, 15)(6)(9)(10)(11, 16)(12, 13)
54. (1)(2, 15, 9)(3, 8, 13)(4, 5, 16)(6, 12, 14)(7, 11, 10)

55. (1)(2, 9, 15)(3, 13, 8)(4, 16, 5)(6, 14, 12)(7, 10, 11)
56. (1)(2, 11)(3, 10)(4)(5)(6, 13)(7, 14)(8, 9)(12, 15)(16)
57. (1)(2)(3, 5)(4, 8)(6, 14)(7)(9, 15)(10, 11)(12)(13, 16)
58. (1)(2, 6)(3, 15)(4)(5)(7, 9)(8, 14)(10, 12)(11, 13)(16)
59. (1)(2)(3, 6)(4, 11)(5, 14)(7)(8, 10)(9, 13)(12)(15, 16)
60. (1)(2)(3, 14)(4, 10)(5, 6)(7)(8, 11)(9, 16)(12)(13, 15)

Irreducible representations of group A5

(1)	1, 46, 52, 53	(7)	7, 32, 45, 51	(8)	8, 24, 36, 59
(13)	6, 14, 37, 47	(14)	5, 15, 27, 58	(15)	16, 33, 54, 57
(16)	11, 13, 44, 48	(19)	4, 20, 42, 60	(20)	2, 21, 31, 56
(21)	22, 25, 49, 50	(22)	3, 19, 35, 55	(40)	23, 34, 40, 43
(41)	12, 26, 30, 41	(46)	10, 18, 28, 39	(52)	9, 17, 29, 38

Underlined elements belong to the congruency nucleus of group **A**5. Group **A**5 has 15 irreducible vector representations, where 12 vectors are the Lorentz ones, time-like; the remaining three vectors form an Euclidean triangle of connection.

Covering G96 group of generating operator of protein synthesis is a holonomy connection group of Lorentz-invariant h-field (Fig. 1.4). Group G96 has a congruency nucleus consisting of 32 elements.

Group G96

1. (1)(2)(3)(4)(5)(6)(7)(8)(9)(10)(11)(12)(13)(14)(15)(16)
2. (1)(2, 16)(3)(4, 13, 6, 15)(5, 12, 7, 14)(8)(9, 11)(10)
3. (1)(2, 16)(3)(4, 14)(5, 13)(6, 12)(7, 15)(8, 10)(9)(11)
4. (1)(2)(3)(4, 7, 6, 5)(8, 10)(9, 11)(12, 13, 14, 15)(16)
5. (1)(2)(3)(4, 6)(5, 7)(8)(9)(10)(11)(12, 14)(13, 15)(16)
6. (1)(2)(3)(4, 5, 6, 7)(8, 10)(9, 11)(12, 15, 14, 13)(16)
7. (1)(2, 16)(3)(4, 12)(5, 15)(6, 14)(7, 13)(8, 10)(9)(11)
8. (1)(2, 16)(3)(4, 15, 6, 13)(5, 14, 7, 12)(8)(9, 11)(10)
9. (1)(2, 5)(3, 9)(4, 14)(6)(7, 10)(8, 15)(11)(12)(13, 16)
10. (1)(2, 15, 16, 7)(3, 9)(4, 12, 14, 6)(5, 10, 13, 8)(11)
11. (1)(2, 13)(3, 9)(4)(5, 16)(6, 12)(7, 8)(10, 15)(11)(14)
12. (1)(2, 6, 10, 14)(3, 11)(4, 16, 12, 8)(5)(7, 15)(9)(13)
13. (1)(2, 10)(3)(4, 12)(5)(6, 14)(7)(8, 16)(9)(11)(13)(15)
14. (1)(2, 14, 10, 6)(3, 11)(4, 8, 12, 16)(5)(7, 15)(9)(13)
15. (1)(2, 6, 8, 4)(3, 11)(5, 7, 13, 15)(9)(10, 14, 16, 12)

16. (1)(2, 8)(3)(4, 6)(5, 13)(7, 15)(9)(10, 16)(11)(12, 14)
17. (1)(2, 4, 8, 6)(3, 11)(5, 15, 13, 7)(9)(10, 12, 16, 14)
18. (1)(2, 12, 8, 14)(3, 11)(4, 16, 6, 10)(5, 15, 13, 7)(9)
19. (1)(2, 14, 8, 12)(3, 11)(4, 10, 6, 16)(5, 7, 13, 15)(9)
20. (1)(2, 4, 10, 12)(3, 11)(5, 13)(6, 16, 14, 8)(7)(9)(15)
21. (1)(2, 12, 10, 4)(3, 11)(5, 13)(6, 8, 14, 16)(7)(9)(15)
22. (1)(2, 8)(3)(4)(5, 15)(6)(7, 13)(9)(10, 16)(11)(12)(14)
23. (1)(2, 10)(3)(4, 14)(5, 7)(6, 12)(8, 16)(9)(11)(13, 15)
24. (1)(2, 13, 16, 5)(3, 9)(4, 6, 14, 12)(7, 10, 15, 8)(11)
25. (1)(2, 5, 16, 13)(3, 9)(4, 12, 14, 6)(7, 8, 15, 10)(11)
26. (1)(2, 7, 16, 15)(3, 9)(4, 6, 14, 12)(5, 8, 13, 10)(11)
27. (1)(2, 8, 16, 10)(3)(4, 5, 12, 15)(6, 7, 14, 13)(9, 11)
28. (1)(2, 10, 16, 8)(3)(4, 13, 12, 7)(5, 6, 15, 14)(9, 11)
29. (1)(2, 15)(3, 9)(4, 14)(5, 8)(6)(7, 16)(10, 13)(11)(12)
30. (1)(2, 7)(3, 9)(4)(5, 10)(6, 12)(8, 13)(11)(14)(15, 16)
31. (1)(2, 8, 16, 10)(3)(4, 7, 12, 13)(5, 14, 15, 6)(9, 11)
32. (1)(2, 10, 16, 8)(3)(4, 15, 12, 5)(6, 13, 14, 7)(9, 11)
33. (1)(2, 5, 4, 16, 13, 14)(3, 9, 11)(6, 8, 7, 12, 10, 15)
34. (1)(2, 4, 13)(3, 11, 9)(5, 16, 14)(6, 7, 10)(8, 12, 15)
35. (1)(2, 13, 4)(3, 9, 11)(5, 14, 16)(6, 10, 7)(8, 15, 12)
36. (1)(2, 14, 13, 16, 4, 5)(3, 11, 9)(6, 15, 10, 12, 7, 8)
37. (1)(2, 5, 10, 7)(3, 9)(4, 6, 14, 12)(8, 15, 16, 13)(11)
38. (1)(2)(3)(4, 15)(5, 12)(6, 13)(7, 14)(8, 10)(9, 11)(16)
39. (1)(2, 13, 10, 15)(3, 9)(4, 12, 14, 6)(5, 8, 7, 16)(11)
40. (1)(2, 6, 15)(3, 11, 9)(4, 5, 10)(7, 16, 12)(8, 14, 13)
41. (1)(2, 15, 6)(3, 9, 11)(4, 10, 5)(7, 12, 16)(8, 13, 14)
42. (1)(2, 16)(3)(4, 7)(5, 6)(8)(9, 11)(10)(12, 13)(14, 15)
43. (1)(2, 7, 10, 5)(3, 9)(4, 12, 14, 6)(8, 13, 16, 15)(11)
44. (1)(2, 6)(3, 11)(4, 8)(5, 13)(7)(9)(10, 14)(12, 16)(15)
45. (1)(2, 5, 12)(3, 9, 11)(4, 8, 7)(6, 16, 13)(10, 15, 14)
46. (1)(2, 12, 5)(3, 11, 9)(4, 7, 8)(6, 13, 16)(10, 14, 15)
47. (1)(2, 7, 14)(3, 9, 11)(4, 16, 15)(5, 6, 8)(10, 13, 12)
48. (1)(2, 14, 7)(3, 11, 9)(4, 15, 16)(5, 8, 6)(10, 12, 13)
49. (1)(2, 13, 12, 16, 5, 6)(3, 9, 11)(4, 10, 7, 14, 8, 15)
50. (1)(2, 12, 15, 16, 6, 7)(3, 11, 9)(4, 13, 10, 14, 5, 8)
51. (1)(2, 16)(3)(4, 5)(6, 7)(8)(9, 11)(10)(12, 15)(13, 14)
52. (1)(2, 15, 10, 13)(3, 9)(4, 6, 14, 12)(5, 16, 7, 8)(11)

53. (1)(2, 4)(3, 11)(5)(6, 8)(7, 15)(9)(10, 12)(13)(14, 16)
54. (1)(2, 6, 5, 16, 12, 13)(3, 11, 9)(4, 15, 8, 14, 7, 10)
55. (1)(2, 15, 14, 16, 7, 4)(3, 9, 11)(5, 12, 8, 13, 6, 10)
56. (1)(2, 4, 7, 16, 14, 15)(3, 11, 9)(5, 10, 6, 13, 8, 12)
57. (1)(2, 7, 6, 16, 15, 12)(3, 9, 11)(4, 8, 5, 14, 10, 13)
58. (1)(2)(3)(4, 13)(5, 14)(6, 15)(7, 12)(8, 10)(9, 11)(16)
59. (1)(2, 12)(3, 11)(4, 10)(5)(6, 16)(7, 15)(8, 14)(9)(13)
60. (1)(2, 4, 5, 8, 12, 7)(3, 11, 9)(6, 15, 16, 14, 13, 10)
61. (1)(2, 10, 16, 8)(3)(4, 5, 14, 13)(6, 7, 12, 15)(9, 11)
62. (1)(2, 8)(3)(4, 14)(5)(6, 12)(7)(9)(10, 16)(11)(13)(15)
63. (1)(2, 5, 14, 8, 7, 6)(3, 9, 11)(4, 10, 15, 12, 16, 13)
64. (1)(2, 14)(3, 11)(4, 16)(5, 13)(6, 10)(7)(8, 12)(9)(15)
65. (1)(2, 8, 16, 10)(3)(4, 13, 14, 5)(6, 15, 12, 7)(9, 11)
66. (1)(2, 12, 7, 10, 14, 13)(3, 11, 9)(4, 5, 16, 6, 15, 8)
67. (1)(2, 7, 8, 13)(3, 9)(4, 14)(5, 16, 15, 10)(6)(11)(12)
68. (1)(2, 7, 12, 8, 5, 4)(3, 9, 11)(6, 10, 13, 14, 16, 15)
69. (1)(2, 6, 13, 10, 4, 7)(3, 11, 9)(5, 8, 14, 15, 16, 12)
70. (1)(2, 10, 16, 8)(3)(4, 7, 14, 15)(5, 12, 13, 6)(9, 11)
71. (1)(2, 4, 15, 10, 6, 5)(3, 11, 9)(7, 8, 12, 13, 16, 14)
72. (1)(2, 8)(3)(4, 12)(5, 7)(6, 14)(9)(10, 16)(11)(13, 15)
73. (1)(2, 12, 16, 6)(3, 11)(4, 8, 14, 10)(5, 7, 13, 15)(9)
74. (1)(2)(3)(4, 14)(5, 15)(6, 12)(7, 13)(8)(9)(10)(11)(16)
75. (1)(2, 14, 16, 4)(3, 11)(5, 15, 13, 7)(6, 8, 12, 10)(9)
76. (1)(2, 5, 6, 10, 15, 4)(3, 9, 11)(7, 14, 16, 13, 12, 8)
77. (1)(2, 13, 8, 7)(3, 9)(4, 14)(5, 10, 15, 16)(6)(11)(12)
78. (1)(2, 13, 14, 10, 7, 12)(3, 9, 11)(4, 8, 15, 6, 16, 5)
79. (1)(2, 6, 7, 8, 14, 5)(3, 11, 9)(4, 13, 16, 12, 15, 10)
80. (1)(2, 15, 8, 5)(3, 9)(4)(6, 12)(7, 10, 13, 16)(11)(14)
81. (1)(2, 15, 12, 10, 5, 14)(3, 9, 11)(4, 16, 7, 6, 8, 13)
82. (1)(2, 12, 13, 8, 4, 15)(3, 11, 9)(5, 10, 14, 7, 16, 6)
83. (1)(2, 8, 16, 10)(3)(4, 15, 14, 7)(5, 6, 13, 12)(9, 11)
84. (1)(2, 14, 15, 8, 6, 13)(3, 11, 9)(4, 7, 10, 12, 5, 16)
85. (1)(2, 10)(3)(4, 6)(5, 15)(7, 13)(8, 16)(9)(11)(12, 14)
86. (1)(2, 6, 16, 12)(3, 11)(4, 10, 14, 8)(5, 15, 13, 7)(9)
87. (1)(2, 16)(3)(4)(5, 7)(6)(8, 10)(9)(11)(12)(13, 15)(14)
88. (1)(2, 4, 16, 14)(3, 11)(5, 7, 13, 15)(6, 10, 12, 8)(9)
89. (1)(2, 13, 6, 8, 15, 14)(3, 9, 11)(4, 16, 5, 12, 10, 7)

90. (1)(2, 5, 8, 15)(3, 9)(4)(6, 12)(7, 16, 13, 10)(11)(14)
91. (1)(2, 14, 5, 10, 12, 15)(3, 11, 9)(4, 13, 8, 6, 7, 16)
92. (1)(2, 10)(3)(4)(5, 13)(6)(7, 15)(8, 16)(9)(11)(12)(14)
93. (1)(2, 7, 4, 10, 13, 6)(3, 9, 11)(5, 12, 16, 15, 14, 8)
94. (1)(2, 16)(3)(4, 6)(5)(7)(8, 10)(9)(11)(12, 14)(13)(15)
95. (1)(2, 15, 4, 8, 13, 12)(3, 9, 11)(5, 6, 16, 7, 14, 10)
96. (1)(2)(3)(4, 12)(5, 13)(6, 14)(7, 15)(8)(9)(10)(11)(16)

Irreducible representations of group G96

	A	**U**	Ψ_U
(1)	1,3,5,7,13,16,	9,11,22,23,29,30,	37,39,43,52

	G,C															
(3)	10,	26	24,	25	62,	67,	72,	74,	77,	80,	85,	87,	90,	92,	94,	96
A (5)	2,	8	4,	6	60,	61,	65,	66,	69,	70,	71,	79,	82,	83,	84,	91
(7)	27,	32	28,	31	34,	36,	38,	40,	42,	46,	48,	50,	51,	54,	56,	58
(13)	12,	14	20,	21	63,	68,	73,	75,	76,	78,	81,	86,	88,	89,	93,	95
(16)	15,	17	18,	19	33,	35,	41,	44,	45,	47,	49,	53,	55,	57,	59,	64

Underlined elements belong to the congruency nucleus of group **G**96. Vector (1) is space-like, vectors (3), (5), (7), (13) and (16) are time-like. The frame contains partially ordered set of 60 elements of group **G**96, which has to be transformed into a group of icosahedron **A**5.

Division of the nucleus representatives of group **G**96 by informational nucleotides:

N = AGAGACACUGUGACGACGCGCUUGCCGGUUCC

N* = ACACAGAGUCUCAGCAGCGCGUUCGGCCUUGG

Nucleotides A and U pair if their self-conjugated representatives commutate. Nucleotides G and C pair if their representatives are conjugated and can be in involution when changing gauge N and N*.

Representatives of nucleotides **A** and **U** determine the bundle of unit $g_s^{\ k}$; representatives of nucleotides **G** and **C** have a type of connection: $\Gamma_{pq}^{\ \ k} = g_s^{\ k}\Gamma_{pq}^{\ \ s}$. Elements of group **G**96 are bound with the elements of group **A**5 by the substitution from symmetric group **S**60: $\mathbf{G}96_b = \omega_b^{\ a}\mathbf{A}5_a$.

Transformation of genetic information

Let us examine the transformation:

$$\mathrm{b}^c = \mathrm{b}^m \mathbf{N} \mathrm{b}^k;\ \mathrm{c} = \mathbf{A}5(m, k),\ \mathrm{b} = \omega\,(\mathrm{a}),\ \mathbf{N} = g, \Gamma$$

If $b^k = 1$, then $b^c = 1$ and b^m is an element of nucleus **G96**. This leads to binding of basic nucleotides and amino acids into one strand, i.e., to generation of additional complexes, for example, of **aminoacyl-AMP**.

Transformation with two nucleotides

$$b^c = N1b^m\,N2\,b^k;\ N1, N2 \subset \mathbf{N}$$

can be realized. Coding nucleotides N1 and N2 bind with each other if pairing vector is a clutching function [8]: $b^k(b^c)^{-1} = \xi = Const$; where obtaining of an inverse element is performed in group **G96**.

The following elements correspond to three paired vectors b^c, b^p and b^t strand of observed nucleotides

$$N1N2N3N4N5N6;$$

strand of virtual vectors

$$b^m, b^k(b^c)^{-1} = \xi, b^c, b^f(b^p)^{-1} = \xi, b^p, b^r(b^t)^{-1} = \xi;$$

vector of two amino acids

$$A_V1 = N1N2N3,\ A_V2 = N4N5N6$$

$$V_{Av1Av2} = b^m, b^c, b^p, b^t;$$

dual vector b^m, b^c, b^p;

factor vector b^k, b^f, b^r.

The proper tensor of torsion

$$\mathbf{T}_{A5} = \begin{vmatrix} c & p & t \\ m & c & p \\ k & f & r \end{vmatrix}$$

corresponds to the matrix of generating operator

$$\Psi_{G96} = \begin{vmatrix} b^c & b^p & b^t \\ b^m & b^c & b^p \\ b^k & b^f & b^r \end{vmatrix}$$

which transfers the vector ξ_{Av1} of an amino acid A_V1 into the vector ξ_{Av2} of an amino acid A_V2.

Selecting the direction of the first vector ξ_{Av1} arbitrarily, each following vector ξ_{Av} binds to the preceding one in an Euclidean space with the relative matrix of turn $\mathbf{T}_{A5}$.

The matrix $\mathbf{T}_{A5}$ consists of the element's numbers of group **A5**. **Identification of matrix elements with the element's numbers of any group determines the conformal infinity of torsion tensor.** Binding of amino acids is performed by a two-component tensor E_{Av1Av2} with connection matrix:

$$E_{Av1Av2} = \begin{vmatrix} m & c \\ p & t \end{vmatrix},$$

which can be identified with the **generating operator's tensor of energy**.

The proper tensor of amino acid energy can be determined as a trace of operator of proper energy in group **A5**:

$$E_{\mathbf{A}} = \mathrm{Tr}\begin{vmatrix} N1 & N2 \\ N2* & N3 \end{vmatrix} = \begin{vmatrix} N1_{A5} & N2_{A5} \\ N2_{A5}{}^{-1} & N3_{A5} \end{vmatrix}$$

where homeomorphism $N1 \to N1_{A5}$ continues until homeomorphism of an inverse element is taken: the inverse element in group **A5** corresponds to the conjugated element in group **G96**. **If the nucleus of homeomorphism is empty, the strand of amino acids loses its biological activity**.

This case corresponds to the binding of two amino acids on a ribosome.

Let us examine the **palindrome transformation** of groups **G96** $\to$ **A5**:

$$b^{\gamma} = N1\, b^{\alpha}\, N2\, b^{\beta}\, N3\, b^{\alpha};\ N1, N2, N3 \subset \mathbf{N};$$

which determines amino acid activation during the time of oscillation of growing polypeptide strand, relatively a ribosome.

The following elements correspond to each physical amino acid A_V = N1N2N3:

proper pairing vector

$$\theta_A = b^{\alpha}, b^{\gamma}, b^{\alpha}(b^{\gamma})^{-1} = \xi$$

and dual vector of pairing

$$\theta_A{}^* = b^{\alpha}, b^{\beta}, b^{\alpha}(b^{\gamma})^{-1} = \xi,$$

which can be united into one vector

$$\mathbf{U}_A = b^{\alpha}, b^{\gamma}, b^{\beta}, b^{\alpha}(b^{\gamma})^{-1} = \xi,$$

as a representation of a disjunctive operator of nucleotide **U**.

Let us determine a torsion tensor of amino acid A_V = N1N2N3 by this way:

$$U_{A96} = \begin{vmatrix} N1 & N3 & N2 \\ N2 & N3 & N1 \\ N3 & N1 & N3* \end{vmatrix} = \begin{vmatrix} A1 \\ A2 \\ A3 \end{vmatrix}.$$

The protein maintains its biological activity if the energy tensor of each amino acid is not empty:

$$E_A = \begin{vmatrix} N1_{A5} & N3_{A5} & N2_{A5} \\ N2_{A5} & N3_{A5} & N1_{A5} \\ N3_{A5} & N1_{A5} & N3_{A5}{}^{-1} \end{vmatrix}.$$

Pairing of amino acids

Basic strand	Unpaired strand	Dual strand
$b^\alpha, b^\gamma, b^\alpha (b^\gamma)^{-1} = \xi$	$b^\alpha, b^\beta, b^\alpha (b^\gamma)^{-1} = \xi$	$b^\alpha, b^\beta, b^\alpha (b^\gamma)^{-1} = \xi$
/		\
$b^\gamma, b^\varepsilon, b^\gamma (b^\varepsilon)^{-1} = \xi$	$b^\gamma, b^\delta, b^\gamma (b^\varepsilon)^{-1} = \xi$	$b^\gamma, b^\alpha, b^\gamma (b^\varepsilon)^{-1} = \xi;$ $\delta = \alpha$
/		\
$b^\varepsilon, b^\mu, b^\varepsilon (b^\mu)^{-1} = \xi$	$b^\varepsilon, b^\sigma, b^\varepsilon (b^\mu)^{-1} = \xi$	$b^\varepsilon, b^\gamma, b^\varepsilon (b^\mu)^{-1} = \xi;$ $\sigma = \gamma$

In order to build an amino acid strand, we arbitrarily determine the tangent vector of first amino acid in the Euclidean space **R**3. Then with regard to this vector we build a vector ξ_A with a matrix of turn $\mathbf{U}_{A96}$. This vector is a tangent vector of second amino acid and so on.

Matrix $\mathbf{U}_{A96}$ – a matrix of the reference point ξ_A – has a strand of homeomorphism:

$$\mathbf{U}_{A96} = \begin{vmatrix} A1 \\ A2 \\ A3 \end{vmatrix} \rightarrow \begin{vmatrix} * \\ A2 \\ A3 \end{vmatrix} \rightarrow \begin{vmatrix} \mathbf{U} & \mathbf{A} & \mathbf{G} \\ & A2 & \\ & & A3 \end{vmatrix} \rightarrow$$

$$\rightarrow \begin{vmatrix} e_m{}^\alpha(x) & 1/2 \cdot \psi_m{}^\alpha(x) & 1/2 \cdot \psi'_{m\alpha'}(x) \\ 0 & \delta_\mu{}^\alpha & 0 \\ 0 & 0 & \delta^{\mu'}{}_{\alpha'}(x) \end{vmatrix} \quad \text{formula (17.2) in ref. [9]}$$

Ψ_U function has four irreducible representations N1N2N3:

$$\Psi_{37} = ggg \quad \sim \quad g_{\alpha}{}^{\gamma} g_{\beta}{}^{\gamma} g_{\alpha}{}^{\gamma} \quad \{8\}$$
$$\Psi_{39} = gg\Gamma, g\Gamma g, \Gamma gg \quad \sim \quad g_{\alpha}{}^{\sigma} g_{\sigma}{}^{\mu} \Gamma_{\mu\beta}{}^{\gamma} \quad \{24\}$$
$$\Psi_{43} = \Gamma\Gamma\Gamma \quad \sim \quad \Gamma_{\alpha\sigma}{}^{\gamma} \Gamma_{\beta\mu}{}^{\sigma} \Gamma_{\alpha\beta}{}^{\mu} \quad \{8\}$$
$$\Psi_{52} = \Gamma\Gamma g, \Gamma g\Gamma, g\Gamma\Gamma \quad \sim \quad \Gamma_{\alpha\mu}{}^{\gamma} \Gamma_{\beta\sigma}{}^{\mu} g_{\alpha}{}^{\sigma} \quad \{24\}$$

(the number of genetic code state is indicated in parentheses).

Indexes of nucleotide representatives are selected by such a way that the trace of function $\boldsymbol{\Psi}_U$ has a type Γ for the amino acids and a type g for the terminators.

Each point of group **A**5 has a set of amino acids suitable for building of Riemann space for the protein envelope. However, there is a logic to select only one random amino acid for each element ω_b^a and to raise the number of layers ω_b^a up to 60!. In this case, the obtained twistor cylindrical space of amino acids is enough for building amino acid strands of any complexity.

Each amino acid vector {A1} = A_V. A1, A2, A3 has a corresponding conjugated vector {A2′} = A_V. A_V, A2′, A3′ of conjugated protein state with the transposed matrix $\mathbf{U}_{A96}$ and the united vector $\{A1\}_6$ = A_V. A1, A2, A3_A2′, A3′.

Tables 1.3-1.8 show the results of calculations for the fixed element ω_b^a of group **S**60:

38, 53, 74, 55, 65, 82, 91, 96, 70, 61, 77, 54, 59, 81, 62, 69, 34, 42, 46, 76, 90, 84, 78, 45, 33, 58, 86, 40, 64, 85,51, 75, 56, 50, 83, 44, 71, 79, 95, 94, 48, 36, 92, 67, 68, 93, 35, 41, 87, 72, 47, 73, 49, 88, 57, 60, 89, 63, 66, 80.

Nucleotide triplet distribution of amino acids is a standard, empirical genetic code.

Table 1.3 Pairing vector of amino acid $\mathbf{U}_A$ in the basis of group **A**5 (fragment)

A5	*1*	*2*	*3*	*4*
1	38.38,38. 1	38.53,53.56	38.74,74.6	38.55,55.53
2	53.53,38. 1	53.82,53.83	53.91,74.8	53.61,55.78
3	74.74,38. 1	74.55,53.71	74.38,74.4	74.53,55.88
4	55.55,38. 1	55.96,53.68	55.70,74.14	55.65,55.86
5	65.65,38. 1	65.77,53.35	65.82,74.39	65.38,55.85
6	82.82,38. 1	82.38,53.90	82.65,74.52	82.77,55.59
7	91.91,38. 1	91.61,53.9	91.53,74.2	91.82,55.13
8	96.96,38. 1	96.74,53.5	96.61,74.31	96.54,55.89
9	70.70,38. 1	70.65,53.7	70.55,74.12	70.96,55.32

Contd.

Table 1.3 Contd

10	61.61,38. 1	61.54,53.26	61.96,74.28	61.74,55.32
11	77.77,38. 1	77.70,53.46	77.54,74.73	77.91,55.53
12	54.54,38. 1	54.91,53.72	54.77,74.86	54.70,55.77
13	59.59,38. 1	59.85,53.75	59.79,74.8	59.84,55.70
14	81.81,38. 1	81.78,53.16	81.51,74.21	81.68,55.3
15	62.62,38. 1	62.64,53.21	62.94,74.13	62.66,55.41
16	69.69,38. 1	69.46,53.87	69.58,74.10	69.95,55.33
17	34.34,38. 1	34.80,53.20	34.63,74.89	34.92,55.60
18	42.42,38. 1	42.94,53.4	42.47,74.53	42.90,55.93
19	46.46,38. 1	46.93,53.76	46.80,74.12	46.87,55.66
20	76.76,38. 1	76.57,53.72	76.72,74.57	76.51,55.88

The length of sequence L_A of amino acids is proportional to the order of a protein group N_P: $L_A = \eta N_P$. Viral envelope of **Aedes DNV** contains a protein consisted of $358 \approx 6 \times 60$ amino acids. Using test genome of **Aedes DNV,** it is possible to build a protein consisting of two strands: $73 + 41 \approx 2 \times 60$ amino acids but with only one intron in its DNA sequence.

Table 1.4 Randomized vector of physical amino acids A_V in the basis of group **A**5 (fragment)

***A**5*	*1*	*2*	*3*	*4*
1	5,29,15	23,30,15	7,14,12	32,30, 5
2	8, 5,24	2,26, 2	4,12, 1	17,28, 2
3	13,13, 6	23, 7,31	6,14,21	7, 1,27
4	23,22,15	14, 3, 1	26,28, 4	8,10, 4
5	8,27, 1	20,11, 1	22,25,12	17,14,12
6	10,29,15	7,18,27	4,19,24	21, 4,12
7	14,24, 6	30,24, 1	6,19,27	25,19, 1
8	1, 7, 2	20,24, 4	14, 9,21	5,32,31
9	3, 6,28	8,23,15	11, 1,10	15, 3,21
10	25,18, 5	30,24,10	22,18,15	13,14,27
11	22,14,15	26, 9, 1	9, 1,14	32,14, 1
12	2, 9,24	10, 8,21	27,13, 1	12, 4,26
13	25, 3,14	17, 9, 2	8,13,12	30,22,12
14	31,18,12	21, 2,12	25, 1, 1	9, 1, 6
15	14,12, 8	13, 1, 3	26,24, 5	28,22, 6
16	27,14,32	25,14,32	28,15,26	22,27,24
17	29,16, 5	3,11,12	1,28,27	16,21,15
18	32,31, 1	7, 1,21	12,13,27	15,19,26
19	24, 9,14	1, 3,12	32,27,27	3,15,12
20	24,15,28	27,20,24	11,23,29	32, 1, 1

Table 1.5 Randomized vector of amino acids $\{A1\}_6 = A_V$. A1, A2, A3_A2′, A3′ in the basis of group **A**5 (fragment)

A5	*1*	*2*	*3*	*4*
1	M.S,*,D_G,Y	L.C,C,V_G,F	T.S,R,D_G,H	L.H,Y,T_N,S
2	Q.R,S,A_G,T	A.G,R,G_G,R	G.E,E,R_K,G	R.R,G,A_G,A
3	N.T,T,Q_P,K	Y.S,T,L_P,M	P.P,P,P_P,P	K.R,R,D_G,N
4	L.C,C,V_G,F	Q.Q,N,T_N,T	R.R,G,A_G,A	R.R,G,A_G,A
5	R.Q,D,T_N,A	V.D,*,R_K,*	S.C,R,V_G,L	P.R,R,A_G,P
6	V.G,W,G_G,C	R.R,G,D_G,D	A.G,R,G_G,R	R.R,G,A_G,A
7	R.P,A,P_P,A	*.*,D,I_N,V	P.R,R,A_G,P	P.H,H,T_N,P
8	K.R,R,D_G,N	G.G,G,G_G,G	L.P,S,P_P,S	T.T,P,Q_P,Q
9	T.S,R,D_G,H	L.R,C,A_G,S	*.*,S,V_G,I	D.A,T,R_P,R
10	R.Q,D,T_N,A	W.W,G,V_G,V	W.W,G,V_G,V	T.S,R,D_G,H
11	S.C,R,V_G,L	L.H,Y,T_N,S	Y.S,T,L_P,M	P.H,H,T_N,P
12	V.G,W,G_G,C	A.A,P,R_P,R	E.E,K,R_K,R	G.A,A,R_P,G
13	H.P,T,P_P,T	L.R,C,A_G,S	Q.R,S,A_G,T	L.C,C,V_G,F
14	R.R,G,A_G,A	R.R,G,A_G,A	Q.Q,N,T_N,T	Y.S,T,L_P,M
15	R.P,A,P_P,A	K.K,K,K_K,K	R.Q,D,T_N,A	V.A,S,R_P,W
16	A.A,P,R_P,R	P.P,P,P_P,P	G.A,A,R_P,G	W.W,G,V_G,V
17	*.*,N,I_N,I	M.S,*,D_G,Y	R.R,G,D_G,D	T.S,R,D_G,H
18	P.H,H,T_N,P	N.T,T,Q_P,K	E.G,R,G_G,S	A.A,P,R_P,R
19	V.A,S,R_P,W	K.R,R,D_G,N	R.R,G,A_G,A	R.R,G,D_G,D
20	G.G,G,G_G,G	G.G,G,G_G,G	F.F,F,F_F,F	Q.Q,N,T_N,T

Basic and dual strands of amino acids coincide and determine the same protein, but are orthogonal geodesics in a local space V4.

Let us introduce the left current of entropy $\mathbf{j}_L$ of basic strand and the right current of entropy $\mathbf{j}_R$ of dual strand:

$$\mathbf{j}_L = \begin{vmatrix} b^\alpha & b^\gamma \\ b^\gamma & b^\varepsilon \end{vmatrix}; \ \mathbf{j}_R = \begin{vmatrix} b^\alpha & b^\beta \\ b^\gamma & b^\alpha \end{vmatrix}$$

Protein doubling occurs when product of entropy currents is proportional to the Lorentz burst:

$$\mathbf{j}_L \mathbf{j}_R = g_{LR} \cdot \boldsymbol{i}^{\perp} = g_{LR} \begin{vmatrix} 1 & \\ & -1 \end{vmatrix}, \qquad (1.1)$$

where g_{LR} is the metric coefficient.

Two modes of protein doubling come from equation (1.1):

1. Spontaneous doubling

$$b^{\varepsilon} = (b^{\alpha})^*;\ b^{\beta} = (b^{\gamma})^*.$$

2. Connected, normal doubling

$$[b^{\alpha}(b^{\gamma})^{-1} = \xi]^2 = 1.$$

Amino acids A1 and A2 induce the palindrome complementary pairing of amino acids among each other or amino acids and their anticodons (equivalents of terminators), because they contain a marked point N3 in the space V4. **Any amino acid pairing, independently of amino acid position in proteins is connected with the entropy transition inside the single protein molecule as well as among different proteins and is caused by the action of charge-exchanging group.**

Table 1.6 Vector set of physical amino acids A_V over the point (20, 3) of group **A5** (288 vectors in total)

23,10,12	3,10,14	1,10,20	22,10,21	25,10, 5
24,10, 7	26,10,22	10,10,23	7,24,12	13,24,14
16,24,20	5,24,21	10,24, 1	26,24, 3	25,24,13
24,24,16	5,25,12	16,25,14	13,25,20	7,25,21
26,25, 1	10,25, 3	24,25,13	25,25,16	22,26,12
1,26,14	3,26,20	23,26,21	24,26, 5	25,26, 7
10,26,22	26,26,23	16, 2,10	23, 2,24	22, 2,25
7, 4,10	1,4,24	3, 4,25	5, 4,26	6, 4, 5
2, 4, 7	8, 4,13	4, 4,16	13, 6,10	22, 6,24
23, 6,25	16, 6,26	6, 6, 1	2, 6, 3	4, 6,22
8, 6,23	5, 8,10	3, 8,24	1, 8,25	7, 8,26
2, 8, 5	6, 8, 7	4, 8,13	8, 8,16	32,27, 5
31,27, 7	7,27, 9	5,27,11	27,27,13	28,27,16
3,27,29	1,27,30	31,28, 5	32,28, 7	5,28, 9
7,28,11	28,28,13	27,28,16	1,28,29	3,28,30
32,31, 1	31,31, 3	13,31, 9	16,31,11	28,31,22
27,31,23	23,31,29	22,31,30	31,32, 1	32,32, 3
16,32, 9	13,32,11	27,32,22	28,32,23	22,32,29
23,32,30	16,12, 2	1,12, 4	13,12, 6	3,12, 8
20,12, 1	14,12, 3	21,12, 5	12,12, 7	22,14, 2
5,14, 4	23,14, 6	7,14, 8	20,14,13	14,14,16
12,14,22	21,14,23	23,20, 2	7,20, 4	22,20, 6

Contd.

Table 1.6 Contd

5,20, 8	14,20,13	20,20,16	21,20,22	12,20,23
13,21, 2	3,21, 4	16,21, 6	1,21, 8	14,21, 1
20,21, 3	12,21, 5	21,21, 7	7,15,27	5,15,28
22,15,31	23,15,32	15,15,13	18,15,16	17,15,22
19,15,23	1,17,27	3,17,28	13,17,31	16,17,32
18,17, 1	15,17, 3	17,17, 5	19,17, 7	5,18,27
7,18,28	23,18,31	22,18,32	18,18,13	15,18,16
19,18,22	17,18,23	3,19,27	1,19,28	16,19,31
13,19,32	15,19, 1	18,19, 3	19,19, 5	17,19, 7
26, 1,10	10, 1,26	6, 1, 4	2, 1, 8	32, 1,27
31, 1,28	14, 1,12	20, 1,21	15, 1,17	18, 1,19
9, 1, 9	11, 1,11	10, 3,10	26, 3,26	2, 3, 4
6, 3, 8	31, 3,27	32, 3,28	20, 3,12	14, 3,21
18, 3,17	15, 3,19	11, 3, 9	9, 3,11	24, 5,10
25, 5,26	2, 5, 2	6, 5, 6	32, 5,31	31, 5,32
21, 5,14	12, 5,20	19, 5,15	17, 5,18	29, 5, 9
30, 5,11	25, 7,10	24, 7,26	6, 7, 2	2, 7, 6
31, 7,31	32, 7,32	12, 7,14	21, 7,20	17, 7,15
19, 7,18	30, 7, 9	29, 7,11	3, 9,15	22, 9,17
1, 9,18	23, 9,19	30, 9, 5	29, 9, 7	9, 9,22
11, 9,23	1,11,15	23,11,17	3,11,18	22,11,19
29,11, 5	30,11, 7	11,11,22	9,11,23	24,13,24
25,13,25	4,13, 4	8,13, 8	28,13,27	27,13,28
14,13,14	20,13,20	18,13,15	15,13,18	30,13,29
29,13,30	25,16,24	24,16,25	8,16, 4	4,16, 8
27,16,27	28,16,28	20,16,14	14,16,20	15,16,15
18,16,18	29,16,29	30,16,30	10,22,24	26,22,25
4,22, 2	8,22, 6	27,22,31	28,22,32	12,22,12
21,22,21	19,22,17	17,22,19	9,22,29	11,22,30
26,23,24	10,23,25	8,23, 2	4,23, 6	28,23,31
27,23,32	21,23,12	12,23,21	17,23,17	19,23,19
11,23,29	9,23,30	16,29,15	7,29,17	13,29,18
5,29,19	9,29, 1	11,29, 3	29,29,13	30,29,16
13,30,15	5,30,17	16,30,18	7,30,19	11,30, 1
9,30, 3	30,30,13	29,30,16		

Table 1.7 Vector set of the amino acids $\{A1\}_6$ over the point (20, 3) of group **A5**

W.W,G,V_G,V	S.T,A,Q_P,E	R.R,G,D_G,D	C.S,A,L_P,V	R.Q,D,T_N,A
G.E,E,R_K,G	R.L,V,S_F,A	G.V,V,C_L,G	R.R,G,D_G,D	S.T,A,Q_P,E
R.R,G,D_G,D	S.T,A,Q_P,E	G.E,E,R_K,G	R.Q,D,T_N,A	R.Q,D,T_N,A
G.E,E,R_K,G	T.S,R,D_G,H	T.T,P,Q_P,Q	T.S,R,D_G,H	T.T,P,Q_P,Q
P.H,H,T_N,P	A.D,Q,R_K,R	A.D,Q,R_K,R	P.H,H,T_N,P	S.C,R,V_G,L
T.T,P,Q_P,Q	T.S,R,D_G,H	S.S,P,L_P,L	A.D,Q,R_K,R	P.H,H,T_N,P
A.V,L,C_L,R	P.L,L,S_F,P	R.R,G,D_G,D	W.W,G,V_G,V	C.S,A,L_P,V
S.T,A,Q_P,E	G.E,E,R_K,G	R.Q,D,T_N,A	R.L,V,S_F,A	G.V,V,C_L,G
R.R,G,D_G,D	R.R,G,D_G,D	S.T,A,Q_P,E	S.T,A,Q_P,E	R.Q,D,T_N,A
G.E,E,R_K,G	R.Q,D,T_N,A	G.E,E,R_K,G	T.S,R,D_G,H	S.C,R,V_G,L
S.S,P,L_P,L	T.T,P,Q_P,Q	P.H,H,T_N,P	A.D,Q,R_K,R	A.V,L,C_L,R
P.L,L,S_F,P	T.S,R,D_G,H	T.S,R,D_G,H	T.T,P,Q_P,Q	T.T,P,Q_P,Q
A.D,Q,R_K,R	P.H,H,T_N,P	A.D,Q,R_K,R	P.H,H,T_N,P	R.Q,D,T_N,A
R.Q,D,T_N,A	S.M,V,Y_L,D	S.M,V,Y_L,D	G.E,E,R_K,G	G.E,E,R_K,G
S.M,V,Y_L,D	S.M,V,Y_L,D	R.Q,D,T_N,A	R.Q,D,T_N,A	S.M,V,Y_L,D
S.M,V,Y_L,D	G.E,E,R_K,G	G.E,E,R_K,G	S.M,V,Y_L,D	S.M,V,Y_L,D
P.H,H,T_N,P	P.H,H,T_N,P	T.I,L,Y_L,H	T.I,L,Y_L,H	A.V,L,C_L,R
A.V,L,C_L,R	S.F,L,F_F,L	S.F,L,F_F,L	P.H,H,T_N,P	P.H,H,T_N,P
T.I,L,Y_L,H	T.I,L,Y_L,H	A.V,L,C_L,R	A.V,L,C_L,R	S.F,L,F_F,L
S.F,L,F_F,L	R.R,G,D_G,D	R.R,G,D_G,D	S.T,A,Q_P,E	S.T,A,Q_P,E
G.E,E,R_K,G	R.Q,D,T_N,A	R.Q,D,T_N,A	G.E,E,R_K,G	S.C,R,V_G,L
T.S,R,D_G,H	S.S,P,L_P,L	T.T,P,Q_P,Q	A.D,Q,R_K,R	P.H,H,T_N,P
A.V,L,C_L,R	P.L,L,S_F,P	W.W,G,V_G,V	R.R,G,D_G,D	C.S,A,L_P,V
S.T,A,Q_P,E	R.Q,D,T_N,A	G.E,E,R_K,G	R.L,V,S_F,A	G.V,V,C_L,G
T.S,R,D_G,H	T.S,R,D_G,H	T.T,P,Q_P,Q	T.T,P,Q_P,Q	P.H,H,T_N,P
A.D,Q,R_K,R	A.D,Q,R_K,R	P.H,H,T_N,P	R.R,G,D_G,D	R.R,G,D_G,D
C.S,A,L_P,V	C.S,A,L_P,V	G.E,E,R_K,G	G.E,E,R_K,G	R.L,V,S_F,A
R.L,V,S_F,A	T.S,R,D_G,H	T.S,R,D_G,H	T.T,P,Q_P,Q	T.T,P,Q_P,Q
A.D,Q,R_K,R	A.D,Q,R_K,R	P.H,H,T_N,P	P.H,H,T_N,P	R.R,G,D_G,D
R.R,G,D_G,D	C.S,A,L_P,V	C.S,A,L_P,V	G.E,E,R_K,G	G.E,E,R_K,G
R.L,V,S_F,A	R.L,V,S_F,A	T.S,R,D_G,H	T.S,R,D_G,H	T.T,P,Q_P,Q
T.T,P,Q_P,Q	A.D,Q,R_K,R	A.D,Q,R_K,R	P.H,H,T_N,P	P.H,H,T_N,P
Q.R,S,A_G,T	D.A,T,R_P,R	Q.R,S,A_G,T	D.A,T,R_P,R	Q.R,S,A_G,T
Q.R,S,A_G,T	Q.R,S,A_G,T	D.A,T,R_P,R	D.A,T,R_P,R	D.A,T,R_P,R
Y.L,I,F_F,I	Y.L,I,F_F,I	E.G,R,G_G,S	H.P,T,P_P,T	E.G,R,G_G,S
H.P,T,P_P,T	Q.R,S,A_G,T	Q.R,S,A_G,T	E.G,R,G_G,S	H.P,T,P_P,T
D.A,T,R_P,R	D.A,T,R_P,R	Y.L,I,F_F,I	Y.L,I,F_F,I	E.G,R,G_G,S
H.P,T,P_P,T	E.G,R,G_G,S	H.P,T,P_P,T	H.P,T,P_P,T	H.P,T,P_P,T
H.P,T,P_P,T	E.G,R,G_G,S	Q.R,S,A_G,T	Q.R,S,A_G,T	Y.L,I,F_F,I
Y.L,I,F_F,I	Q.R,S,A_G,T	D.A,T,R_P,R	Q.R,S,A_G,T	D.A,T,R_P,R

Contd.

Table 1.7 Contd

H.P,T,P_P,T	H.P,T,P_P,T	D.A,T,R_P,R	Q.R,S,A_G,T	Q.R,S,A_G,T
Q.R,S,A_G,T	Y.L,I,F_F,I	Y.L,I,F_F,I	M.S,*,D_G,Y	F.S,S,L_P,L
M.S,*,D_G,Y	F.S,S,L_P,L	L.Y,Y,I_N,L	L.Y,Y,I_N,L	F.F,F,F_F,F
F.F,F,F_F,F	M.S,*,D_G,Y	F.S,S,L_P,L	M.S,*,D_G,Y	F.S,S,L_P,L
L.Y,Y,I_N,L	L.Y,Y,I_N,L	F.F,F,F_F,F	F.F,F,F_F,F	E.G,R,G_G,S
H.P,T,P_P,T	E.G,R,G_G,S	H.P,T,P_P,T	E.G,R,G_G,S	E.G,R,G_G,S
H.P,T,P_P,T	E.G,R,G_G,S	E.G,R,G_G,S	E.G,R,G_G,S	Y.L,I,F_F,I
Y.L,I,F_F,I	Q.R,S,A_G,T	D.A,T,R_P,R	Q.R,S,A_G,T	D.A,T,R_P,R
E.G,R,G_G,S	E.G,R,G_G,S	D.A,T,R_P,R	Q.R,S,A_G,T	E.G,R,G_G,S
E.G,R,G_G,S	Y.L,I,F_F,I	Y.L,I,F_F,I	V.G,W,G_G,C	L.P,S,P_P,S
V.G,W,G_G,C	L.P,S,P_P,S	V.A,S,R_P,W	V.A,S,R_P,W	V.G,W,G_G,C
L.P,S,P_P,S	L.P,S,P_P,S	L.P,S,P_P,S	F.F,F,F_F,F	F.F,F,F_F,F
L.R,C,A_G,S	V.A,S,R_P,W	L.R,C,A_G,S	V.A,S,R_P,W	V.A,S,R_P,W
V.A,S,R_P,W	L.R,C,A_G,S	V.A,S,R_P,W	L.P,S,P_P,S	L.P,S,P_P,S
F.F,F,F_F,F	F.F,F,F_F,F	M.S,*,D_G,Y	I.T,S,Q_P,*	M.S,*,D_G,Y
I.T,S,Q_P,*	L.Y,Y,I_N,L	L.Y,Y,I_N,L	L.Y,Y,I_N,L	L.Y,Y,I_N,L
M.S,*,D_G,Y	I.T,S,Q_P,*	M.S,*,D_G,Y	I.T,S,Q_P,*	L.Y,Y,I_N,L
L.Y,Y,I_N,L	L.Y,Y,I_N,L	L.Y,Y,I_N,L		

We deduce from Table 1.7 that each amino acid A1 has quite a large set of rotation matrices. Vector of amino acids $\{A1\}_6$ is unambiguously determined by vector {A1} and it does not depend on the point of group **A5** and element $\omega_b^{\ a}$.

Cyclic indexes of group **G96** are equal to 1, 2, 3, 4, 6; they coincide with the observed cyclic indexes of the nucleotide triplet distribution of amino acids. The volume of basic amino acids coincides with the sub-nucleus connectedness of group **G96**. The volume of the set of virtual amino acids over points of group **A5** takes values (9, 11, 13) × 32 and covers 393 amino acids of factor EF-Tu [10]. It is possible, therefore, to suppose that **group G96 is really a minimum group of generating operator of protein synthesis.**

The structure of the carrier particle of tRNA amino acids completely reflects basic elements of the genome transformation into the protein. The constant vector **CCA** reflects the mode of transformation of the vector space's generating operator: $\mathbf{C}\, b^{\alpha}\, \mathbf{A}\, b^{\beta}\, \mathbf{C}\, b^{\alpha} = 1$. A minor and three basic loops of tRNA cloverleaf determine the vector of amino acid {A1}. Varying of the number of nucleotides' representatives is compensated by the presence of modified nucleotides in tRNA. Function $\mathbf{\Psi}_U$ is also located in tRNA and corresponds to one of four amino acid types.

In the activated protein, when each amino acid turns from state A_V into one of the states A1 or A3, the polypeptide strand becomes free from the group A5 and becomes a gauge vector field with variable quantity of vectors of the tangent charge space.

Let Q(A) – dimension of vector space (set of vectors A_V) for each amino acid in a polypeptide strand of the complete set be $Q(A)_{max} = (9, 11, 13) \times 32$. Then Q(A1), Q(A3) is the accumulated charge of an amino acid A1 and A3 and a **strand of amino acids becomes a memory element with charge bond of address space A_V.** When protein grades into the conjugated state, charges Q(A2′) and $Q(A_V)$ correspond to the storage of information Q(A): charge Q(A2′) maintains during protein doubling, charge $Q(A_V)$ determines basic, non-active, degenerated state of protein. When the normal period of doubling is completed, protein returns to one of the states A1, A3. Vector field $Q(A1, A3, A2', A_V)$ induces wave movements of the polypeptide strand, generation of pairs A3_A2′, as well as interactions of protein during the exchange of information Q(A).

Dual state of protein is described by three vectors of amino acids: vector

$\{A3'\} = A_V$. A1, A2, A3′, vector conjugated to the preceding one,

$\{A1'\} = A_V$. A1′, A2′, A3′ with the formation of united vector,

$\{A3'\}_6 = A_V$. A1, A2_A1′, A2′, A3′ and self-conjugated factor vector.

$\{A_V\} = A_V$. A1, A2′, A3′. Vector field Q(A1, A2, A1′, A2′) induces interaction of paired amino acids with the formation of connection triangle

A3_A2′, A2_A1′, A_V_A3′.

The set of vectors $\{A1\}_6$ is invariant in respect of all 2^{10} gauge transformations **G**, **C** → **C**, **G** of vectors **N**, **N***.

If group **A5** is replaced by an Abelian group $C(3) \times C(4) \times C(5)$, then the vector field of amino acids $\{A1\}_6$ will not change; the set of vectors $\{A1\}_6$ over points of group $C(3) \times C(4) \times C(5)$, will have the same $(9, 11, 13) \times 32$ values as in the group **A5**. This fact proves the universality of the built gauge field of amino acids. Let us note that P, F, G and K amino acids can remain invariant relatively to the biological activation of protein: the protein consisting of these amino acids determines the basic physical structure of a protein bundle with transition into the biologically active form. Homeomorphism P → **C**, F → **U**, G → **G**, K → **A** is a reflection of protein in the RNA strand.

The permutation of the amino acids A1 and A3 in matrix $\mathbf{U}_{A96}$ induces protein change from the state {A1} into the conjugated state {A2′}.

Performed mathematical transformations of genome have an analogue in quantum mechanics.

Table 1.8 The volume of the amino acid $\{A1\}_6$ vectors set over the points of groups A5/C(3) × C(4) × C(5) in combined basis (fragment)

	1	2	3	4	5	6
1	352/352	288/288	352/352	288/288	352/352	288/288
2	288/288	416/352	288/352	416/288	352/416	416/352
3	416/416	352/288	416/288	352/352	352/352	288/288
4	416/416	352/352	416/352	416/416	288/352	288/288
5	288/288	288/416	288/288	352/352	416/416	416/416
6	416/416	288/288	352/288	416/288	352/288	416/288
7	352/352	288/352	352/352	288/416	288/416	288/352
8	416/416	352/416	416/288	288/352	352/416	288/288
9	288/288	288/352	288/288	288/352	352/416	352/352
10	288/288	352/352	352/416	288/288	288/416	288/288
11	288/288	288/352	288/416	288/288	352/416	416/288
12	352/352	352/416	352/288	416/352	352/352	352/352
13	288/288	352/416	288/352	288/352	288/288	416/352
14	416/416	352/416	416/288	288/288	352/288	352/352
15	352/352	352/352	352/416	288/288	352/352	352/288
16	416/416	352/416	416/288	288/416	416/352	288/416
17	416/416	288/416	352/352	288/288	288/416	352/288
18	352/352	416/416	288/288	352/352	352/352	288/288
19	352/352	352/288	416/352	288/416	288/288	288/288
20	352/352	416/352	288/352	352/288	416/352	416/352

Schrödinger equation

$$(H_0 + V^\wedge)\Psi = E\Psi$$

can be written as

$$(E–H_0)\Psi = V^\wedge\Psi;$$

$$\Psi = \{(E–H_0)^{-1} V^\wedge (E–H_0)^{-1} V^\wedge (E–H_0)^{-1} V^\wedge\}\Psi = E_{HV}{}^\wedge\Psi$$

Let $\Gamma^\wedge = V^\wedge V^\wedge V^\wedge$ be the connection operator with a set of irreducible proper components $\{\Gamma\}$. Then, $E_{HV}{}^\wedge = (E–H_0)^{-1}\lambda$ and $H_0\Psi = (E–\lambda)\Psi$.

The operator of interaction V^ congruently transforms the spectrum of tangent Gilbert space of Ψ vectors into Riemann space with connection {Γ}.

In the simplest case $i\partial\Psi/\partial t = -\lambda\Psi$ and bundle Ψ by group SU(2) is introduced. **RNA is indeed a flow of central informational charges, which directly change in time the state vector of protein synthesis generating operator.**

Quantized Fields of Genetic Code

Chapter 2

There are many works devoted to the elucidation of genetic code nature. Let us review these works [11, 12], where possible evolution of the genetic code on the basis of Lie groups has been examined. The set of 20 observed amino acids of genetic code has been determined by means of consecutive bundle of one hypothetic amino acid with the exclusion of those amino acid transformations, which lead to deviation from the observed genetic code. The main disadvantage of such formal constructions is that amino acid methyonine and terminators do not differ from other amino acids.

Degeneracy of genetic code is a hard problem. It is clear that RNA information does not have to be lost while replacement of the alphabet consisted of four nucleotides by the alphabet, which consists of 20 amino acids.

The problem of protein structure doubling is very complex. All attempts to explain the phenomenon of protein doubling in physical four-dimensional space were not successful.

In order to resolve these problems, we propose an approach based on the theory of quantized fields in Riemann space. Continual field models in projected spaces [13, 14] as well as the theory of Yang-Mills field in four-dimensional space-time [1] are based on presentation of the charge space as a space of Lie group states. We introduce a conformal field of amino acids, which removes degeneracy of the genetic code. This field is also a gauge field but with an external metric induced by mixed gauge.

The theory of Penrose twistors [15] is used for the presentation of conformal field, as it has an infinite transformation group, which is not Lie group. We use group **G**96 for the presentation of symmetry group of protein synthesis generating operator as a modular presentation of supersymmetric algebra of conformal field transformations. The power of these transformations is equal to 60!.

The transition of amino acids from physical base state to biological state occurs during the oscillation of growing amino acid strand with respect to the ribosome. Each amino acid in protein determines vector gauge field $\{A1\}_6$ connected to the triplet of coding RNA nucleotides $N1N2N3 \subset N$ in the following way:

$$\begin{aligned} \{A1\}_6 &= Av, A1, A2, A3\_A2', A3', \\ Av &= N1N2N3,\ A1 = N1N3N2,\ A2 = N2N3N1, \\ A3 &= N3N1N3^*,\ A2' = N3N3N1,\ A3' = N2N1N3^*. \end{aligned}$$

Conjugated nucleotides are determined via the basic ones:

$$\mathbf{A^* = A, U^* = U, G^* = C, C^* = G.}$$

The rotation matrix of the amino acids pairing vector ξ_A for the physically observed amino acid A_V can be determined by following the formula:

$$\mathbf{U}_A = \begin{vmatrix} N1N3N2 \\ N2N3N1 \\ N3N1N3^* \end{vmatrix}.$$

Generating operator of protein synthesis has a holonomy connection group **G**96 of Lorentz-invariant h-field of Hadamard matrix H16, which contains congruency nucleus consisting of 32 elements – representatives of basic nucleotides:

$$\begin{aligned} \mathbf{N} &= \mathbf{AGAGACACUGUGACGACGCGCUUGCCGGUUCC} \\ \mathbf{N^*} &= \mathbf{ACACAGAGUCUCAGCAGCGCGUUCGGCCUUGG} \end{aligned}$$

and determines the spine metric 2^+2^- of RNA charge space.

Conformal-infinite matrix $\mathbf{U}_A$ consisting of numbers, of group **G**96, is reduced to a finite group O(3) by orthogonalization according to Gram-Schmidt method, where the first vector of orthogonalization process is that vector (A1, A2 or A3), which determines the state of activated amino acid A_V. If det $\mathbf{U}_A = 0$ then $\mathbf{U}_A$ is determined by an ideal: for vector $A_V = 1, 1, 1$

$$\mathbf{U}_A(1,1,1) = \begin{vmatrix} 1 & 1 & 1 \\ -1 & 0 & 1 \\ 1 & -2 & 1 \end{vmatrix},$$

the cosine of rotation angle of three-dimensional Euclidean space around the proper vector of matrix $\mathbf{U}_A$ is equal to $\cos\delta = 7.200718 \cdot 10^3$; $3/\cos\delta = 416.62$. Tangent charge space of amino acid consists of vectors set $\{A1\}_6$ over the point of protein group. This set can contain 288, 352 or 416 vectors. The accumulated charge of amino acid in activated state is determined from the same multitude.

Gauge field of amino acids is a representation of genetic code. **The state of each amino acid is determined by 'quantum numbers,' which are amino acids themselves**. Local change of the genetic code basis is compensated by the appearance of gauge amino acid field. After the normal ordering of amino acids in a new basis, we obtain an initial code.

Breaking of the genetic code gauge invariance leads to the appearance of compensating conformal field of amino acids (CFAA) with an external metric.

In order to describe the properties of CFAA we introduced a six-dimensional physical space with a signature $4^+ 2^-$, which corresponds to CFAA signature and possesses doubling properties. Binominal set $C^k{}_6 = \{1, 6, 15, 20, 15, 6, 1\}$ contains 20 connection elements, which are identified with observed amino acids of genetic code and together with terminator* form a basis of finite projected plane of the order four. Six-dimensional physical space is a bundle on the finite four-dimensional projected space with signature $3^+ 1^-$ with enclosed finite projected plane of the fourth order of self-acting scalar field. Basic set of amino acids is formed as a result of self-acting of scalar vector. This fixes an observed gauge of the genetic code amino acids and conformal field retains its 'freezing.'

In this chapter we prove that conformal field of amino acids is bundle over physical "hard" protein basis and is **identical to a gravitational field with a spin of 5/2**. Mechanism of CFAA doubling, metaphase spindle of CFAA of the cell and colour space of genetic code were built; and the Yang-Mills fields of amino acid bundle were examined.

Table 2.1 represents CFAA in observed gauge of genetic code as a vector field $\{A1\}_6$ over arbitrary protein point.

Table 2.1 Conformal field of amino acids of genetic code

A.D,Q,R_K,R(**GCA**)	A.G,R,G_G,R(**GCG**)
_A,	_A,
A.A,P,R_P,R(**GCC**)	A.V,L,C_L,R(**GCU**)
_A,	_A,
C.S,A,L_P,V(**UGC**)	C.L,V,F_F,V(**UGU**)
_W,	_W,
D.A,T,R_P,R(**GAC**)	D.V,M,C_L,S(**GAU**)

Contd.

Table 2.1 Contd

_E,	_E,
E.E,K,R_K,R(**GAA**)	E.G,R,G_G,S(**GAG**)
_E,	_E,
F.S,S,L_P,L(**UUC**)	F.F,F,F_F,F(**UUU**)
_F,	_F,
G.E,E,R_K,G(**GGA**)	G.G,G,G_G,G(**GGG**)
_G,	_G,
G.A,A,R_P,G(**GGC**)	G.V,V,C_L,G(**GGU**)
_G,	_G,
H.P,T,P_P,T(**CAC**)	H.L,I,S_F,T(**CAU**)
_Q,	_Q,
I.N,*,K_K,*(**AUA**)	I.T,S,Q_P,*(**AUC**)
_I,	_I,
I.I,L,Y_L,Y(**AUU**)	
_I,	
K.K,K,K_K,K(**AAA**)	K.R,R,D_G,N(**AAG**)
_K,	_K,
L.H,Y,T_N,S(**CUA**)	L.R,C,A_G,S(**CUG**)
_L,	_L,
L.P,S,P_P,S(**CUC**)	L.L,F,S_F,S(**CUU**)
_L,	_L,
L.Y,Y,I_N,L(**UUA**)	L.C,C,V_G,F(**UUG**)
_F,	_F,
M.S,*,D_G,Y(**AUG**)	
_I,	
N.T,T,Q_P,K(**AAC**)	N.I,I,Y_L,N(**AAU**)
_K,	_K,
P.H,H,T_N,P(**CCA**)	P.R,R,A_G,P(**CCG**)
_P,	_P,
P.P,P,P_P,P(**CCC**)	P.L,L,S_F,P(**CCU**)
_P,	_P,
Q.Q,N,T_N,T(**CAA**)	Q.R,S,A_G,T(**CAG**)
_Q,	_Q,
R.K,E,K_K,E(**AGA**)	R.R,G,D_G,D(**AGG**)
_R,	_R,
R.Q,D,T_N,A(**CGA**)	R.R,G,A_G,A(**CGG**)
_R,	_R,
R.P,A,P_P,A(**CGC**)	R.L,V,S_F,A(**CGU**)
_R,	_R,

Contd.

Table 2.1 Contd

S.T,A,Q_P,E(**AGC**)	S.M,V,Y_L,D(**AGU**)
_R,	_R,
S.Y,H,I_N,L(**UCA**)	S.C,R,V_G,L(**UCG**)
_S,	_S,
S.S,P,L_P,L(**UCC**)	S.F,L,F_F,L(**UCU**)
_S,	_S,
T.N,Q,K_K,Q(**ACA**)	T.S,R,D_G,H(**ACG**)
_T,	_T,
T.T,P,Q_P,Q(**ACC**)	T.I,L,Y_L,H(**ACU**)
_T,	_T,
V.D,*,R_K,*(**GUA**)	V.G,W,G_G,C(**GUG**)
_V,	_V,
V.A,S,R_P,W(**GUC**)	V.V,L,C_L,C(**GUU**)
_V,	_V,
W.W,G,V_G,V(**UGG**)	
_W,	
Y.S,T,L_P,M(**UAC**)	Y.L,I,F_F,I(**UAU**)
_*,	_*,
.,N,I_N,I(**UAA**)	*.*,S,V_G,I(**UAG**)
_*,	_*,
.,D,I_N,V(**UGA**)	
_W,	

Note: The lower under-line corresponds to dual conformal field of amino acids.

Table 2.2 represents charge space characteristics of amino acids on the basis of total protonic charge of each amino acid's radical. The marked point of proline is equivalent to an antiproton. It follows from the fact that pirrolidine ring of proline $(-\mathbf{NH}-\mathbf{CH_2}-\mathbf{CH_2}-\mathbf{CH_2}-\mathbf{CH_2}-)$ has a negative charge –1(mod M(41)). Linear strand $(\mathbf{CH_2}-\mathbf{CH_2}-\mathbf{CH}=)$ is a radical of proline.

Chemical elements **C, H, O, N, S**, and antiproton can play a role of basic generators of group Sp(6) in a six-dimensional space with metric $4^+\,2^-$, because they commutate in pairs. These generators are determined on the basis of representation of group Sp(6) in a charge-exchanging group, which is isomorphic to a Braid group [3] and they perform six functions: connection, spiralization, autoinclusion, Lorentz turns, parallel transition and terminator. We identified the functions of some generators.

Table 2.2 Characteristics of amino acid charge space.

Amino acid Am	*Chemical formula of the radical*					*Protonic charge of the radical*			
						Q_p	Q_p (mod 4) = h(Am)	Q_p (mod 32)	Q_p (mod 6)
A–ALANINE	1C	3H				9	+1	9	3
C–CYSTEINE	1C	3H			1S	25	+1	25	1
D–ASPARTIC ACID	2C	3H	2O			31	−1	31	1
E–GLUTAMIC ACID	3C	5H	2O			39	−1	7	3
F–PHENYLALANINE	7C	7H				49	+1	17	1
G-GLYCINE		1H				1	+1	1	1
H–HISTIDINE	4C	5H		2N		43	−1	11	1
I–ISOLEUCINE	4C	9H				33	+1	1	3
K–LYSINE	4C	10H		1N		41	+1	9	5
L–LEUCINE	4C	9H				33	+1	1	3
M–METHIONINE	3C	7H			1S	41	+1	9	5
N–ASPARAGINE	2C	4H	1O	1N		31	−1	31	1
*P–PROLINE	3C	5H				23	−1	23	5
Q–GLUTAMINE	3C	6H	1O	1N		39	−1	7	3
R–ARGININE	4C	10H		3N		55	−1	23	1
S–SERINE	1C	3H	1O			17	+1	17	5
T–THREONINE	2C	5H	1O			25	+1	25	1
V–VALINE	3C	7H				25	+1	25	1
W–TRYPTOPHAN	9C	12H		1N		73	+1	9	1
Y–TYROSINE	7C	7H	1O			57	+1	25	3

Note: * – marked point of terminator

Amino acid signature $13^+ 7^-$ (mod 4), $Q_p(x) = -1$

Signature: 1(mod 6) = $6^+ 4^-$; $Q_p(3+) = 6(4^+ 2^-)$; $Q_p(5^-) = 4(3^+ 1^-)$

Charge increment + 10 is required for normal division (mod 32)

Groups A5 and G96

Gauge transformations of amino acids A2 → A3: N1N2N3 → N3N1N3* reflects CFAA on vectors of group **A5** (Table 2.3). Vector (–K, F, P) determines the connection of the group (15-parametric Lie group). Let {g…} be a set of vectors of protein symmetry group contained in DNA. Then, a form of group **A5** is used for transferring this group upon amino acid sequence in protein.

The following order types are congruency nuclei of group **A5**: ω_1) palindromes – parallel transfer of group: g_c = N1g_a N2g_b N3g_a with connection nucleus **ACA** of tRNA acceptor site **ACCA**; ω_2) involution and

Table 2.3 Irreducible representations of group A5

Am = A(A3_A2′, A3′_A2)			
A_G,T_S(**CAG**)	A_G,S_C(**CUG**)	A_G,A_G(**CGG**)	A_G,P_R(**CCG**)
C_L,S_M(**GAU**)	C_L,C_L(**GUU**)	C_L,G_V(**GGU**)	C_L,R_L(**GCU**)
D_G,N_R(**AAG**)	D_G,Y_*(**AUG**)	D_G,D_G(**AGG**)	D_G,H_R(**ACG**)
F_F,I_I(**UAU**)	F_F,F_F(**UUU**)	F_F,V_V(**UGU**)	F_F,L_L(**UCU**)
G_G,S_R(**GAG**)	G_G,C_W(**GUG**)	G_G,G_G(**GGG**)	G_G,R_R(**GCG**)
I_N,I_N(**UAA**)	I_N,L_Y(**UUA**)	I_N,V_D(**UGA**)	I_N,L_H(**UCA**)
Ψ'(**U**) :			
K_K,K_K(**AAA**)	K_K,*_*(**AUA**)	K_K,E_E(**AGA**)	K_K,Q_Q(**ACA**)
L_P,M_T(**UAC**)	L_P,L_S(**UUC**)	L_P,V_A(**UGC**)	L_P,L_P(**UCC**)
P_P,T_T(**CAC**)	P_P,S_S(**CUC**)	P_P,A_A(**CGC**)	P_P,P_P(**CCC**)
Q_P,K_T(**AAC**)	Q_P,*_S(**AUC**)	Q_P,E_A(**AGC**)	Q_P,Q_P(**ACC**)
R_P,R_T(**GAC**)	R_P,W_S(**GUC**)	R_P,G_A(**GGC**)	R_P,R_P(**GCC**)
S_F,T_I(**CAU**)	S_F,S_F(**CUU**)	S_F,A_V(**CGU**)	S_F,P_L(**CCU**)
T_N,T_N(**CAA**)	T_N,S_Y(**CUA**)	T_N,A_D(**CGA**)	T_N,P_H(**CCA**)
V_G,I_S(**UAG**)	V_G,F_C(**UUG**)	V_G,V_G(**UGG**)	V_G,L_R(**UCG**)
Y_L,N_I(**AAU**)	Y_L,Y_L(**AUU**)	Y_L,D_V(**AGU**)	Y_L,H_L(**ACU**)
Ψ(**A**) :			
R_K,R_K(**GAA**)	R_K,*_*(**GUA**)	R_K,G_E(**GGA**)	R_K,R_Q(**GCA**)

Note: Underlined elements belong to the congruency nucleus of group **A5**

twisting of group: $g_c = N1g_b\ N2g_a\ N3g_a$ with connection nuclei **ACC** and **CAA** of tRNA acceptor site **ACCA**. In this case DNA is being transcribed with a shift.

Palindrome transformation of groups **G96** → **A5**

$$b^{\gamma} = N1\, b^{\alpha}\ N2\, b^{\beta}\ N3\, b^{\alpha};\ \{ b^{\alpha}, b^{\beta}, b^{\gamma} \} \subset \mathbf{G96}$$

determines amino acid activation during oscillation of growing polypeptide strand with regard to the ribosome.

Vector $\Psi(\mathbf{A})$ is a normal divisor of group **G96**; it is pushed out by simple group **A5** and attaches to group **G96** (Table 2.4) after involuntary transformation $\Psi(\mathbf{A}) \rightarrow \Psi'(\mathbf{A})$:

$$Am = Am(A3\_A2', A3'\_A2) \rightarrow Am' = Am(A2'\_A3, A3'\_A2).$$

Group **G96** converts vector $\Psi'(\mathbf{U})$ into vector $\Psi(\mathbf{U})$.

Table 2.4 Irreducible representations of group G96. Gauge transformation: A3 → A2′: N1N2N3 → N3N3N1; Am′ = Am(A3_A2′, A3′_A2)

<1> <2>	<1> <3>	<1> <4>	<1> <5>
G_A,T_S(**CAG**)	G_A,S_C(**CUG**)	G_A,A_G(**CGG**)	G_A,P_R(**CCG**)
G_D,N_R(**AAG**)	G_D,Y_*(**AUG**)	G_D,D_G(**AGG**)	G_D,H_R(**ACG**)
G_G,S_R(**GAG**)	G_G,C_W(**GUG**)	G_G,G_G(**GGG**)	G_G,R_R(**GCG**)
G_V,I_S(**UAG**)	G_V,F_C(**UUG**)	G_V,V_G(**UGG**)	G_V,L_R(**UCG**)
F_F,I_I(**UAU**)	F_F,F_F(**UUU**)	F_F,V_V(**UGU**)	F_F,L_L(**UCU**)
F_S,T_I(**CAU**)	F_S,S_F(**CUU**)	F_S,A_V(**CGU**)	F_S,P_L(**CCU**)
L_C,S_M(**GAU**)	L_C,C_L(**GUU**)	L_C,G_V(**GGU**)	L_C,R_L(**GCU**)
L_Y,N_I(**AAU**)	L_Y,Y_L(**AUU**)	L_Y,D_V(**AGU**)	L_Y,H_L(**ACU**)
¥(U) :			
K_K,K_K(**AAA**)	K_K,*_*(**AUA**)	K_K,E_E(**AGA**)	K_K,Q_Q(**ACA**)
¥′(A) :			
K_R,R_K(**GAA**)	K_R,*_*(**GUA**)	K_R,G_E(**GGA**)	K_R,R_Q(**GCA**)
N_I,I_N(**UAA**)	N_I,L_Y(**UUA**)	N_I,V_D(**UGA**)	N_I,L_H(**UCA**)
N_T,T_N(**CAA**)	N_T,S_Y(**CUA**)	N_T,A_D(**CGA**)	N_T,P_H(**CCA**)
P_L,M_T(**UAC**)	P_L,L_S(**UUC**)	P_L,V_A(**UGC**)	P_L,L_P(**UCC**)
P_P,T_T(**CAC**)	P_P,S_S(**CUC**)	P_P,A_A(**CGC**)	P_P,P_P(**CCC**)
P_Q,K_T(**AAC**)	P_Q,*_S(**AUC**)	P_Q,E_A(**AGC**)	P_Q,Q_P(**ACC**)
P_R,R_T(**GAC**)	P_R,W_S(**GUC**)	P_R,G_A(**GGC**)	P_R,R_P(**GCC**)

Let us examine reflection <2> + <4> ↔ <3> + <5>:

(A):	P	H	G	L	Q	I	T	M	N	S	K	R
F(A):	↑	↑	↓	↑	↑	↓	↓	↓	↓	↓	↓	↓
(B):	A	D	R	V	E	*	S	W	Y	L	C	F

Arrows are determined in such a way so that h(A) ± h(B) = 0(mod 4).

The representation F(A) determines the superalgebra **AF** of numbers of amino acid occupation numbers:

AF =	A	C	D	E	F	G	H	I	K	L	M	N	P	Q	R	S	T	V	W	Y	*
	1	1	1	1	1	1	1	1	1	2	1	1	1	1	2	2	1	1	1	1	1

All amino acids obey Fermi statistics. Amino acids L, R, S are in two conjugated spin states. Let us calculate the commutator of fields **C**(Am, Am˜), where Am˜ is dual field, in algebra **AF**, using marked points (field charges): L.R,C,A_G,S and S.M,V_R,L,D.

We have: **C**(Am, Am˜) = 0

Normal CFAA divisor – metaphase spindle has a form (Fig. 2.1):

1. Points R, L, D bind points S, M, V and form three-dimensional system of reference.
2. Transformation S.M,V_R,L,D → S.M,V,Y_L,D doubles protein.
3. Divergence poles are D, G.
4. Division is necessary because the transition L.R,C,A_G,S → L.R,C_L,G,S is prohibited in algebra **AF**.

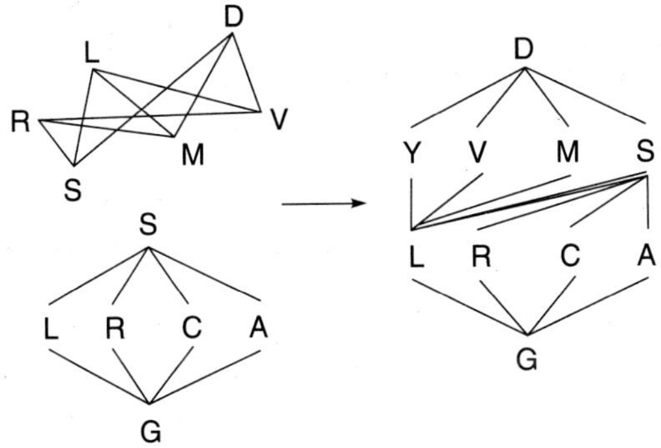

Fig. 2.1 CFAA metaphase spindle. Vectors MD, ML and MR are orthogonal in point M.

Let us calculate the symmetrical differences of protonic charges of CFAA normal divisor:

$$Q_pL = Q_{pLR}(L + S + G + R + C + A) \rightarrow Q_p(3\mathbf{N}) = 21.$$
$$Q_pR = Q_{pRL}(L + S + D + M + V + Y) \rightarrow Q_p(9\mathbf{C}\ 7\mathbf{H}\ 3\mathbf{O}) = 85.$$
$$Q_{\Psi Am} = Q_pR - Q_pL = 64.$$

Here, indexes L, R determine left and right parts of division spindle. Hence, **metaphase spindle structure is formed as a result of space congruence of amino acid states.**

Matrix 'α'

Let us determine the amino acid sign according to the formula:

$$Q_4(Am) = h(Am)Q_p(Am);\ Q_4(*) = -1.$$

Let us build a sign matrix 'α' (Fig. 2.2), according to Table 2.1, of the involutive conformity (Av, A1) → A2 taking into account the signs of amino acids and then select the corresponding covering Hadamard matrix H20 (Fig. 2.3). The sum of diagonal elements of the matrix 'α' by H20 matrix is equal to:

$$Q_4(-4P + 2K + 2F + 2L - N + G) + 5^+1^- + Q_4(-N + S - D) = \mathbf{GF}(83).$$

A1:	A	C	D	E	F	G	H	I	K	L	M	N	P	Q	R	S	T	V	W	Y	*
Av:																					
A	-P	0	-Q	0	0	-R	0	0	0	0	0	0	0	0	0	0	0	L	0	0	
C	0	0	0	0	0	0	0	0	0	V	0	0	0	0	0	A	0	0	0	0	
D	T	0	0	0	0	0	0	0	0	0	0	0	0	0	0	0	0	M	0	0	
E	0	0	0	K	0	-R	0	0	0	0	0	0	0	0	0	0	0	0	0	0	
F	0	0	0	0	F	0	0	0	0	0	0	0	0	0	0	S	0	0	0	0	
G	A	0	0	-E	0	G	0	0	0	0	0	0	0	0	0	0	0	V	0	0	
H	0	0	0	0	0	0	0	0	0	I	0	0	T	0	0	0	0	0	0	0	
I	0	0	0	0	0	0	0	L	0	0	0	-*	0	0	0	0	S	0	0	0	
K	0	0	0	0	0	0	0	0	K	0	0	0	0	0	-R	0	0	0	0	0	
L	0	C	0	0	0	0	Y	0	0	F	0	0	S	0	C	0	0	0	0	Y	
M	0	0	0	0	0	0	0	0	0	0	0	0	0	0	0	-*	0	0	0	0	
N	0	0	0	0	0	0	0	I	0	0	0	0	0	0	0	0	T	0	0	0	
P	0	0	0	0	0	0	-H	0	0	L	0	0	-P	0	-R	0	0	0	0	0	
Q	0	0	0	0	0	0	0	0	0	0	0	0	0	-N	S	0	0	0	0	0	
R	0	0	0	0	0	0	0	0	-E	V	0	0	A	-D	-**G**	0	0	0	0	0	
S	0	-R	0	0	L	0	0	0	0	0	V	0	0	0	0	-P	A	0	0	-H	
T	0	0	0	0	0	0	0	L	0	0	0	-Q	0	0	0	-R	-P	0	0	0	
V	S	0	-*	0	0	W	0	0	0	0	0	0	0	0	0	0	0	L	0	0	
W	0	0	0	0	0	0	0	0	0	0	0	0	0	0	0	0	0	0	G	0	
Y	0	0	0	0	0	0	0	0	0	I	0	0	0	0	0	T	0	0	0	0	
*																					-N+S-D

Fig. 2.2 Matrix 'α' of involutive correspondence (Av, A1) → A2, signed. Point R.R is marked point GG = –G(mod 3*).

-	1	**-**	1	1	**-**	-	-	1	1	-	1	1	1	-	-	-	**1**	-	1
-	1	1	-	-	-	-	1	-	**1**	-	1	-	-	-	**1**	1	1	1	1
1	-	1	1	-	-	1	-	1	1	-	-	1	-	-	-	1	**1**	1	-
1	1	1	**1**	-	**-**	-	-	-	1	1	1	-	1	1	-	1	-	-	-
-	-	1	1	**1**	-	1	1	1	1	-	1	-	-	1	**1**	-	-	-	-
1	1	-	**-**	-	**1**	1	1	1	1	1	1	-	-	-	-	-	**1**	-	-
-	-	-	-	-	-	1	-	-	**1**	1	1	**1**	1	1	1	-	1	1	-
1	1	-	-	1	-	1	**1**	-	1	**-**	-	1	1	-	1	**1**	-	-	-
1	-	-	1	-	-	-	1	**1**	1	1	-	-	1	**-**	1	-	-	1	1
1	**1**	1	1	1	1	**1**	1	1	**1**	1	1	**1**	1	**1**	1	1	1	1	**1**
1	-	1	-	1	-	-	1	-	1	1	-	1	-	1	**-**	-	1	-	1
-	-	-	1	-	1	1	**1**	-	1	-	-	-	1	1	-	**1**	1	-	1
-	-	1	-	1	1	**-**	-	1	**1**	1	-	**-**	1	**-**	1	1	1	-	-
-	1	-	-	1	-	1	-	1	1	1	-	-	**-**	**1**	-	1	-	1	1
-	1	1	1	-	1	1	-	**-**	**1**	1	-	**1**	**-**	**-**	1	-	-	-	1
-	**-**	-	1	**1**	1	-	1	-	1	**1**	1	1	-	-	**-**	**1**	-	1	**-**
-	1	1	-	-	1	-	**1**	1	1	-	**-**	1	1	1	**-**	**-**	-	1	-
1	1	**-**	1	1	**1**	-	-	-	1	-	-	-	-	1	1	-	**1**	1	-
1	-	1	-	1	1	1	-	-	1	-	1	-	1	-	-	-	-	**1**	1
1	-	-	-	-	1	-	-	1	**1**	-	1	1	-	1	**1**	1	-	-	1

Fig. 2.3 Covering Hadamard matrix of signed matrix 'α'. Signature: $J_L = J_R = 13^+ 7^-$.

If we do not take into account the amino acid signs when reflecting (Av, A1) → A2, then covering Hadamard matrix is canonically transformed and we will have a signature of $J_L = 17^+ 3^-$, $J_R = 8^+ 12^-$ (Fig. 2.3a). In this case

1	1	**1**	-	-	**1**	-	-	1	-	-	1	1	-	1	1	-	**1**	-	-
1	1	1	-	1	-	-	-	-	**1**	1	-	-	-	1	**1**	1	-	-	1
1	1	1	1	1	-	1	-	-	-	1	1	-	-	-	-	-	**1**	1	-
1	1	-	**1**	-	**1**	1	-	-	-	-	1	-	1	-	1	1	-	-	1
1	-	-	-	**1**	1	-	1	-	1	-	1	-	-	-	**1**	-	1	1	1
1	-	-	**1**	1	**1**	-	-	1	-	1	-	1	-	-	-	1	**1**	-	1
1	-	1	-	-	-	1	-	1	**1**	-	1	**1**	-	-	-	1	-	1	1
1	1	-	-	-	-	-	**1**	1	1	1	**1**	-	1	-	-	**1**	1	-	-
1	1	-	-	1	1	1	1	**1**	-	-	-	-	-	**1**	-	1	-	1	-
1	**1**	-	1	-	-	**1**	1	-	**1**	-	-	**1**	-	**1**	-	-	1	-	**1**
1	-	-	-	-	-	1	-	1	-	1	-	-	1	1	**1**	-	1	1	1
1	-	1	-	1	-	1	**1**	-	-	-	-	1	1	-	1	**1**	1	-	-
1	-	-	-	1	1	**1**	-	-	**1**	1	1	**1**	1	**1**	-	-	-	-	-
1	-	1	1	1	-	-	1	1	-	-	1	-	**1**	**1**	-	-	-	-	1
1	1	-	1	1	-	-	-	**1**	**1**	-	-	**1**	**1**	**-**	1	-	-	1	-
1	**1**	1	1	**1**	1	1	1	1	1	**1**	1	1	1	1	**1**	**1**	1	1	**1**
1	-	-	1	-	-	-	**1**	-	-	1	**1**	1	-	1	**1**	**1**	-	1	-
1	-	**1**	1	-	**1**	-	-	-	1	-	-	-	1	1	-	1	**1**	1	-
1	1	1	-	-	1	-	1	-	-	1	-	1	1	-	-	-	-	**1**	1
1	-	1	1	-	1	1	1	1	**1**	1	-	-	-	-	**1**	-	-	-	-

Fig. 2.3a Covering Hadamard matrix of unsigned matrix 'α' but with preservation of marked point. Signature: $J_L = 17^+ 3^-$, $J_R = 8^+ 12^-$.

$$Q_p(4P + 2K + 2F + 2L + N + G + *) + 4^+ 2^- - Q_4(-N + S - D) = 416.$$

Let us pass the gauge from charge A2(N2N3N1) to charge A3(N2N3N2*) in the matrix 'α' with metric –1, +1, –1 of three-dimensional space. The result of this compression transformation will be that 20 amino acids will be turned to 15 amino acids of congruency nuclei of group **A**5.

Obtained sub-nucleus of the connection (nucleolus) has a charge:

$$Q_4(A + C - D + F + G + I + K + L - P - Q - R + S + T + V + Y) = \mathbf{GF}(167).$$

Factor metric of CFAA six-dimensional space is equal to:

$$Q_4(-E, -H, +M, -N, +W, -*) = 0.$$

This transformation induces transfers with $\delta Q_4(Am) = 0$:

$$Q_4\{K \to N\} = -72; Q_4\{R \to S\} = +72; Q_4\{M \to I\} = -8; Q_4\{E \to D\} = +8;$$
$$Q_4\{2L \to 2F\} = +32; Q_4\{2H \to 2Q\} = +8; Q_4\{W \to C\} = -48;$$
$$Q_4\{C \to S*\} = +8; Q_4\{Y \to N*\} = -26; Q_4\{Y \leftarrow *D\} = +26$$

and transformation of three-dimensional metric is

$$Q_4\{-N, S, -D \to K, -R, -D\} = 0.$$

The three-dimensional metric has a coloured charge $CL = Q_4\{-N + S - D\} = -45$.

Field H20 is gauge-invariant. It means that the CFAA compression transformation is reversible and hence, can be transformed into radiation. CFAA radiation process is CFAA doubling.

The splitting of terminators causes splitting of double point Y. We can identify terminators N* and S*. The obtained new point (*) expands metric –1, +1, –1 to Lorentz metric +1, –1, –1, –1. Let us attach points (*) and *D to CFAA.

H20 field has a holonomy connection group with a volume of 72 elements with cyclic indexes 1, 2, 3, 4, 6. Congruence nucleus has 22 involute elements. Let us transform Hadamard matrix of field H20 by involute twistor transformation π_ω to such a form that congruence nucleus of the field contain cycles of length six. Then derivative group $G_a{}^\wedge G_b{}^\wedge = g_a g_b g_a$ coincides with the initial one and will contain two immovable points (*), *D, and congruence nucleus consists of 20 elements, which determines the 20 observed amino acids.

Field H20 is associated with tRNA. This is a superconductive field, because the density charge matrix transported by this field is proportional to the field itself [3].

Let us represent matrix 'α' on three planes of charges –N, +S and –D. As a result, genetic code forms a compacted coloured space of dodecahedron states with the distinguished point GG and the supporting colours N, S, and D. Let us transform matrix 'α' into a free set of amino acids with preservation of marked points:

$$Q_4\{-4P - 2Q - 6R + 6L + 4V + 4A + 4T + 1M + 2K + 2F$$
$$+ 5S - S^* - 2E + 2G - G + 3I + 2C + 2Y - 2H - 1N$$
$$+1N^* - 1D - 1{}^*D + 1W\} = 2\mathbf{GF}(167)$$

Let us divide the coloured charge CL = –45 into three ideals using field **GF**(5) of parallel transfer of matrix 'α' onto two fields **GF**(167) by a colourless module 15 of factor metric of CFAA field compression. We will then replace the termination mark* of factor metric by marked point –G. As a result of division, we obtain sets:

$$2Q_4(A + C - D + F + G + I + K + L - P - Q - R$$
$$+ S + T + V + Y) = 2\mathbf{GF}(167).$$
$$Q_4(-2P - 2R + 4L + 2V + 2A + 2T + 2S^* - E$$
$$+ I - H + N^* - {}^*D) = 0$$
$$Q_4(-E - H + M - N + W - G) = 0$$

It has to be noted, that

$$\mathbf{GF}(13) \times \mathbf{GF}(2^5) = 2[\mathbf{GF}(167) + \mathbf{GF}(41)] = \mathbf{GF}(83) \times 5^{+} - 1^{-} = 416.$$

Parallel transfer of amino acids into a new system of reference by means of reflection of basic RNA nucleotides into basic amino acids **A** → K, **U** → F, **C** → P, **G** → G generates self-dual field of bundle of amino acids G, F, P, K.

The field of parallel transfer has a maximum nucleus, which consists of 15 elements:

$$Q_4(A + C - D - E + F + G - H + K + L - N - P - R + S + T + V) = \mathbf{GF}(3).$$

with factor metric of six-dimensional self-dual CFAA space:

$$Q_4(+I, +M, -Q, +W, +Y, -{*}) = 164.$$

Yang-Mills fields

Let us divide the field **GF**(83) into sets, +**GF** and –**GF** by attaching Galois fields located in a root subspace to the first set and Galois fields from a root supplement to the second one. We will then determine the bundle of amino acids in field **GF**(83):

+**GF**{Am}: –1 2* 5 8* 13 19 32** 43 47 53 67 71 73 79

↑

Am : G A S P C,T,V D,N I,L E,Q K,M H F R Y W

Q_p\{Am\}: 1 9 17 23 25 31 33 39 41 43 49 55 57 73

↓

–**GF**{Am}: 1 3 4* 7 9 11 16* 17 23 25 27 29 31 37 41 49 59 61 64* 81

according to amino acid protonic charge and lexicographic order.

Let us calculate total protonic charge of amino acids at each charge strand:

$$Q_p(Am) = Q_p\{G(1) + A(9) + S(17) + P(23) + C(25) + T(25) + V(25) + D(31) + N(31)$$

$$+ I(33) + L(33) + E(39) + Q(39) + K(41) + M(41) + H(43) + F(49) + R(55) + Y(57) + W(73)\}$$

$$Q_p\mathbf{GF}(Am)^{+} = Q_p\{G(1) + A(2^*) + S(5) + P(8^*) + 3V(13) + 2N(19)$$

$$+ 2L(32^*) + Q(43) + M(47) + H(53) + F(67) + R(71) + Y(73) + W(79)\}$$

$$Q_p\mathbf{GF}(Am)^{-} = Q_p\{G(1) + A(3) + S(4^*) + P(7) + C(9) + T(11) + V(16^*) + D(17) + N(23) +$$

$$+ I(25) + L(27) + E(29) + Q(31) + K(37) + M(41) + H(49) + F(59) + R(61) + Y(64^*) + W(81)\}$$

Entropy is determined by neutral charges:

$$\Omega^+ = Q_p\{A(2^*) + P(8^*) + L(32^{**})\};$$
$$\Omega^- = Q_p\{S(4^*) + V(16^*) + Y(64^*)\}.$$

Sign sums are determined using character h(Am):

$$Q_4(\mathrm{Am}) = Q_4\{G + A + S + P - C + T + V + D - N - I + L + E - Q - K + M + H - F + R - Y + W + {}^*\}$$
$$Q_4\mathbf{GF}(\mathrm{Am})^- = Q_4\{G + A - P - C + T - D + N - I + L - E + Q - K + M + H + F - R + W + {}^*\}$$
$$Q_4\mathbf{GF}(\mathrm{Am})^+ = Q_4\{G + S + 3V - 2N - 2Q - 2M + H - F - R + Y - W\}.$$

Postfix sign notation is used in the case of $Q_4(\mathrm{Am})$ and $Q_4\mathbf{GF}(\mathrm{Am})^-$, and prefix is used for $Q_4\mathbf{GF}(\mathrm{Am})^+$.

$$Q_p\mathbf{GF}(\mathrm{Am})^- - Q_p(\mathrm{Am}) = Q_4\mathbf{GF}(\mathrm{Am})^+ + Q_4(\mathrm{Am}) + \mathbf{GF}(3^*)$$

The entropy production:

$$Q_p\mathbf{GF}(\mathrm{Am})^+ - Q_p(\mathrm{Am}) = \Omega^+ - \Omega^-.$$

Obtained correlations can be formulated as an **equivalency principle**:

(a) **Any connection is equal to some space-time Lorentz-interval;**

(b) **Any field is determined by the sum of some connection and entropy production**:

$$\begin{aligned} Q_p(\mathrm{Am}) &= Q_p\mathbf{GF}(\mathrm{Am})^- - Q_4\mathrm{GF}(\mathrm{Am})^+ - Q_4(\mathrm{Am}) - \mathbf{GF}(3^*). \\ Q_p\mathbf{GF}(\mathrm{Am})^+ &= Q_p(\mathrm{Am}) + \Omega^+ - \Omega^-. \\ Q_4\mathbf{GF}(\mathrm{Am})^- - Q_4(\mathrm{Am}) &= P(4); \Omega^- = 4P(4), \end{aligned}$$

where P(4) is dimension of the basis of finite projected plane of the order four.

Covering field of genetic code: $\mathbf{GF}(83) = \Omega^{-*}$.

Induced Yang-Mills field is determined through potential $YM_4(\mathrm{Am})$ of transfer $Q_p(\mathrm{Am})$ into $Q_p\mathbf{GF}(\mathrm{Am})$:

$$YM_4(\mathrm{Am})^+ = Q_4\{G + S + C + T + V + D - N - E - Q - K - M - H + F - R - Y + W - {}^*\}$$
$$YM_4(\mathrm{Am})^- = Q_4\{G - A - P + C - T + D - N + I - L + E - Q + K + M + H - F + R + W\}.$$

Postfix and prefix sign notations are used in $YM_4(\mathrm{Am})^+$ and $YM_4(\mathrm{Am})^-$ respectively.

$$-YM_4(\mathrm{Am})^+ - Q_4(\mathrm{Am}) - {}^* = \mathbf{GF}(41) = M - \text{Yang-Mills field}.$$

$$YM_4(Am)^- - Q_4(Am) = \phi = Am(6) - \text{conformal field of amino acids.}$$

And, finally:

$$\phi - YM_4(Am)^+ - YM_4(Am)^- - * = M,$$

that proves the existence of the amino acids' conformal field.

Also note that $Q_p(C) - Q_p\mathbf{GF}(C)^- = Q_p\{^{32}\mathbf{S}_{16}\}$. Chemical element **S** is the shift generator in charge space of amino acids inducing Yang-Mills field (connection).

Let us examine the parallel transfer of space-time interval:

$$\{M(41)^+, H(43), W(73)^+\}^+ \xrightarrow{G(1)} \{K(41)^+, A(9)^+, F(49)^+\}.$$
$$Q_p\{K, A, F\} - Q_4\{M, H, W\}^+ = \mathbf{GF}(3^{3^*}).$$

The diagram of amino acid transfers is shown on Fig. 2.4. Here, $r_m = \varphi[(\varphi\{Q_p(Am)\}]$ is root subspace Am; φ is Euler theoretical numerical function. Sign r_m determines transfer of straight$^{(+)}$ and additional$^{(-)}$ roots.

Each amino acid has as many inputs as the quantity of amino acid derivatives. Chemical element **H** = G(1) is generator of parallel transfer of Lorentz-interval.

Let us now examine the amino acid bundle in fields **GF**(83) and **GF**(167). The sign of amino acid in these fields will be determined by the following way:

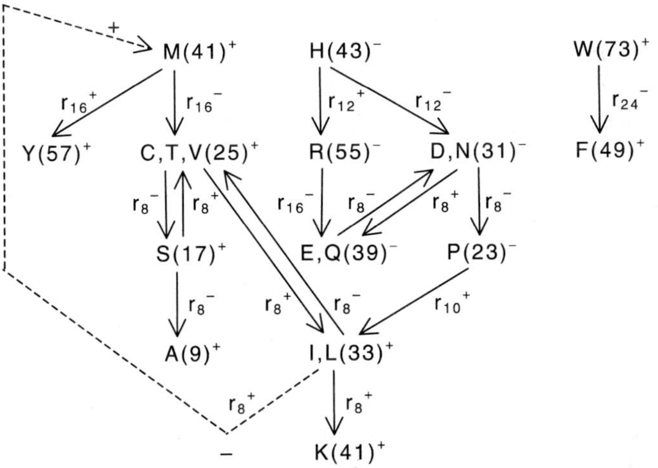

Fig. 2.4 Parallel transfer of Lorentz-interval

$$\chi(\mathrm{Am}) = \begin{cases} +1, \mathrm{Q_p}\{\mathrm{Am}\} \subset \mathrm{r}(\mathbf{GF}) \\ -1, \mathrm{Q_p}\{\mathrm{Am}\} \not\subset \mathrm{r}(\mathbf{GF}), \end{cases}$$

where r(**GF**) is root subset of Galois field.

Potentials of Yang-Mills field in these fields are equal to:

$$\mathrm{YM83(Am)} = \mathrm{Q}_\chi\{-\mathrm{G} - \mathrm{A} - \mathrm{S} - \mathrm{P} - \mathrm{C} - \mathrm{D} - \mathrm{I} + \mathrm{E} - \mathrm{K} + \mathrm{H} - \mathrm{F} + \mathrm{R} + \mathrm{Y} + \mathrm{W} - 3^*\} = \mathrm{M}(41).$$

$$\mathrm{YM167(Am)} = \mathrm{Q}_\chi\{-\mathrm{G} - \mathrm{A} + \mathrm{S} + \mathrm{P} - \mathrm{C} - \mathrm{D} - \mathrm{I} + \mathrm{E} + \mathrm{K} + \mathrm{H} - \mathrm{F} + \mathrm{R} - \mathrm{Y} + \mathrm{W} + 3^*\} = \mathbf{GF}(83).$$

The field of parallel transfer **GF**(3*) is introduced here.

We shall prove **that amino acid M(41) is the first amino acid of field ϕ.**

For this, let us examine the segment Q_p(Am). According to Staudt theorem [16], in a projected correspondence established between the points of the same straight line, it cannot be more than two double points if this correspondence it not identical. On segment Q_p(Am), there are seven double points of finite projected plane P(2), which is reflected on finite projected plane P(4) in a six-dimensional space with signature $2^+ 4^-$: 2P(4) in Q_p**GF**$(Am)^+$, 4P(4) in Q_p**GF**$(Am)^-$. Double points are located consecutively on segment C(25) – M(41). Infinitely remote point* corresponds to I, L point. Reflection $Q_p(Am) \rightarrow Q_p$**GF**$(Am)^+$ has two immovable points H and W in accordance with Staudt theorem. Reflection $Q_p(G\ldots K) \rightarrow Q_p$**GF**$(G\ldots K)^-$ is a gauge invariant: seven double points of reflection correspond to the seven double points of a derivative Q_p(Am), which is concentrated on segment C(25) – K(41).

$Q_p(M\ldots W) \rightarrow Q_p$**GF**$(M\ldots W)^-$ representation is the projected correspondence.

Let us perform a shift of Q_p(Am) to the right on 32**. Point G will be turned into I, L point. We will use reflection

$$Q_p(G\ldots K) \rightarrow Q_p\mathbf{GF}(G\ldots K)^-$$

as a connection. For example,

$$Q_p(S) \rightarrow Q_p(F); Q_p(D) \rightarrow Q_p\mathbf{GF}(D)^- \rightarrow Q_p(F).$$

Points $Q_p(I)$, $Q_p(K)$, $Q_p(F)$, $Q_p(R)$, $Q_p(Y)$, $Q_p(W)$ are new double points. Point $Q_p(A)$ will be turned into $Q_p(H)$ and hence, is not a double point. Double point $Q_p(E, Q)$ is projected into Q_p**GF**$(E, Q)^+$ and determines the start of movement Q_p(Am) relatively to Q_p**GF**$(D)^+$ and Q_p**GF**$(D)^-$ (σ-factor) because finite projected plane of the order six does not exist. Yang-Mills field M(41) shifts to W(73). The termination of movement Q_p(Am) is caused by reflection of relative leader W(73) and the formation of triple point $Q_p(C, T, V) = 3\,Q_p(C)$: $2(I + M) - W = 3C$. Taking into account the identity of constriction $I + M - 3C = *$, we obtain $W - I - * = M$, i.e., mark M is in position 73. It is clear that points M and I are equivalent relatively to W leader.

Now transfer the $Q_p(W)$ leader into point $Q_p\mathbf{GF}(W)^- = Q^p\mathbf{GF}(M)^-$. Then point $Q_p(M)$ will turn into point $Q^p\mathbf{GF}(M)^-$ and will become the first point of an amino acid set as the first point followed after three terminators in $Q_p\mathbf{GF}(Am)^-$. To do it, let us make a shift of set $Q_p(Am)$ by 8*. Point $Q_p(I, L)$ will turn into point $Q_p(M)$; segment $Q_p(A\ldots W) = 64^*$ is a track of new leader mark $Q^p\mathbf{GF}(M)$. Point 64* is located in interval –20 from the first mark $Q^p\mathbf{GF}(M)^-$:

$$\mathbf{GF}(81) = \mathbf{GF}(61) + \mathbf{GF}(17) - \mathbf{GF}(3^*).$$

Thus, terminator field and leader mark $Q_p(W)$, which directly precede Shine-Dalgarno sequence [10] are located before the first mark $Q^p\mathbf{GF}(M)^-$; and this is experimentally demonstrated.

During central projection of sets

$$Q_p(Am) \to Q_p\mathbf{GF}(Am)^-$$

from pole $Q_p\mathbf{GF}(G)^+$, point $Q_p(K, M)$ is projected first. Interval $Q_p(G\ldots K)$ is transformed into two intervals

$$Q_p\mathbf{GF}(G\ldots K)^- \text{ and } Q_p\mathbf{GF}(M\ldots W)^-.$$

Infinitivally-remote point $Q_p(I, L)$ is projected in 64*. Point $Q_p(N)$ is projected in **GF**(61), point $Q_p(D)$ is transferred into connection **GF**(17) and we come to the previous case.

Let us place field ϕ in position A2. Then K will turn into R or stay immovable; M will turn into termination point *. Note that point $Q_p(K, M)$ is projected in $Q_p\mathbf{GF}(W)^-$ from point 2*. Hence, M is the first point ϕ, which was asserted.

External differentiation of set $Q_p(Am)$ determining all 64 amino acid states occurs on interval $Q_p(A\ldots W)$ in field **GF**(167), which contains 50 Galois fields necessary for it.

Reduplication of field M(41)

Let us examine the duplication of field M(41) by matrix 'α', and represent reflection $(A_V, A1) \to A2$ in form of 2×2 matrices of field M(41):

$$\begin{matrix} A_V & (A_V, A1) \\ -(A1, A_V) & A1 \end{matrix}$$

in two gauges (Tables 2.5 and 2.6).

Table 2.5 Field 2S(17) of module Q(35) of matrix 'α'. Reflection (Av, A1) → A2(N2N3N1)

A	–P	A	–Q	A	–R	A	L	C	V	C	A
P	A	–T	D	–A	G	–S	V	–C	L	R	S
				Q (39)							
D	M	E	K	E	–R	F	F	R	–G	G	G
D*	V	–K	E	E	G	–F	F	G	R	–G	G
H	I	H	T	I	L	I	N*	I	S		
–Y	L	H	P	–L	I	–I	N	–L	T		
K	K	K	–R	L	Y	L	F	L	S	L	C
–K	K	E	R	–I	Y	–F	L	–L	P	–V	R
M	–S*	P	–P	P	–R	N	T	Q	–N	Q	S
–V	S	P	P	–A	R	Q	T	N	Q	D	R
S	–P	S	A	S	–H	T	–P	V	L	W	G
P	S	R	T	–T	Y	P	T	–L	V	–G	W

$$K(37) = \begin{vmatrix} G & V \\ -W & V \end{vmatrix} \quad M(41) = \begin{vmatrix} F & S \\ -L & S \end{vmatrix}$$

Table 2.6 Module Q(39) when compressing CFAA. Reflection (Av, A1) → A3(N2N1N2*)

A	–P	A	–Q	A	–R	A	L	C	V	C	A
P	A	–T	D	–A	G	–S	V	–C	L	R	S
D	I	E	K	E	–R	F	F	R	–G	G	G
–Y	V	–K	E	D	G	–F	F	G	R	–G	G
G	V	H	I	H	T	–I	F	I	Y	I	S
–C	V	–Y	L	H	P	–F	–I	–I	N	–L	T
K	K	K	–R	L	Y	L	F	L	S	L	C
–K	K	D	R	–I	Y	–F	L	–L	P	–V	R
M	C	P	–P	P	–R	N	T	Q	–K	Q	–R
–V	S	P	P	–A	R	Q	T	K	Q	D	R
S	–P	S	A	S	–Q	T	–P	V	F	W	G
P	S	R	T	–T	Y	P	T	–F	V	–G	W
–1	K	1	–R	–1	–D						
–K	–1	–R	1	D	–1						

$$M(41) = \begin{vmatrix} F & S \\ -L & S \end{vmatrix}$$

Field M(41) is compressed to the module Q(35) passing through factor-marks with determinant Q(**GF**(M)) = 0. But the mark Q(39) remains in module 2S(17) in force of hyperbolic metric in space of terminators. In module 2S(17), terminators are projected on matrix 'α'.

There is only one factor-mark in module Q(39), but splitting of pairs (I, I), (L, L) is introduced. A hole in module Q(35) is filled with module Q(39). Factors-marks M(41), M(41), K(37) are field of charges N*, D* and –S*. Charge Q(39) mixes the set 2S(17) with module Q(39). After mixing, reduplication of field M(41) occurs by restoration of terminators during CFAA expanding.

CFAA expanding is induced by transfer of

$$\mathrm{K}(41) \rightarrow \mathbf{GF}\{\mathrm{K}(41)\}^{-} = \mathbf{K}(37).$$

Reduplication of M(41) field:

$$\mathrm{Q}(35) + \phi + \mathrm{Q}(39) = 2\mathrm{M}(41)$$

K(37)

induces representation of entropy: $\mathrm{Q}(35) + \mathrm{Q}(39) = \Omega^{+}$.

Thus, mixing of modules Q(35) and Q(39) is necessary in force of preservation of charge Q(39) in module Q(35) and, hence, by the return of reduplicated field M(41) into the initial gauge by matrix 'α'.

Factorization of charge Q(39) leads to the appearance of second hole in module Q(35); both holes are divergence poles of fields M(41). Factor-marks of connection are transformed into the Lorentz-interval

$$\mathbf{K}(37) - \mathrm{Q}(39) + 2\mathrm{M}(41) = \mathrm{Q}^{\mathrm{P}}\mathbf{GF}(\mathrm{M})^{-} - \mathrm{Q}^{\mathrm{P}}\mathbf{GF}(\mathrm{G})^{-}$$

of pole movement.

Left and right matrices of group $\mathrm{O}(2)_{\mathrm{L}}$, $\mathrm{O}(2)_{\mathrm{R}}$ are determined as left and right entropy currents of basic and dual amino acid strands [4] and are related to ordinal type $\omega_1 = \omega_{\mathrm{L}} \cup \omega_{\mathrm{R}}$; while field **GF**(41) also contains matrices of group GL(2, R) of superentropy ordinal type ω_2 [fields $\mathbf{GF}(3^2)$, $\mathbf{GF}(5^2)$].

Remember that the gauge transformation of field A(x) is introduced in the following way:

$$\mathrm{A}_{\mu}(x) \rightarrow \omega(x)\,\mathrm{A}(x)\,\omega(x)^{*-1} + \partial_{\mu}\,\omega(x)\,\omega(x)^{*-1},$$

where $\omega(x)^*$ is a matrix conjugated to matrix $\omega(x)$. Under the differentiation sign, there is a congruency nucleus of holonomy connection group of matrix $\omega(x)$. The permutation of matrix $\omega(x)$ conjugation and conversion operations determines four principal transformation modes. If $\omega(x)^* = \omega(x)^{-1}$ then such a gauge transformation is twistor transformation.

Conformal field ϕ is transformed in the following way:

$$\phi_\mu \to \omega_1 \phi_\mu \omega_1 + \partial_\mu \omega_2.$$

Also note that four particles Q_p(C, T, V), Q_p(D, N), Q_p(I, L) and Q_p(K, M) can be identified with β, α, α, β' particles of RNA-polymerase enzyme; multitude Q_p**GF**$(Am)^-$ can be identified with (–) DNA strand and Q_p**GF**$(Am)^+$ – with (+) DNA strand. Then tracking of enzyme movement (ordinal type of which is equal to ω_1) along (–) DNA strand is equal to 64* and is identified with RNA.

Interval Q_p**GF**$(W)^+$ – Q_p**GF**$(GG)^+$ is the same as an interval of mark M(41) movement, therefore, RNA is homeomorphous to (+) DNA strand. Ribosome subunits are identified with ordinal type ω_2.

Reduplication of conformal field

Table 2.7 contains CFAA in a mixed gauge: components of field Av.A1 are left in the calibration of nucleotides triplet N1N2N3, components of field A2, A3_A2′, A3′ are determined in gauge N3N1N2. Metric 4^+2^- is introduced by this transformation in a six-dimensional vector space of CFAA. The equality of amino acid potentials Am(N1N2N3) = Am(N3N1N2) induces preservation of the charges:

$$\mathbf{Q1}(\mathrm{N1N2N3}) = \mathbf{Q4}(\mathrm{N3N2N1}),$$

$$\mathbf{Q2}(\mathrm{N1N3N2}) = \mathbf{Q3}(\mathrm{N3N1N2}).$$

Let us study the CFAA twistor gauge:

$$\phi\,[X_A = \mathrm{Am}(3),\ Y_A = \mathbf{Am(3)}],\ \text{where}$$

$$\mathrm{Am}(3) = \begin{vmatrix} \mathrm{N1N2N3} \\ \mathrm{N1N3N2} \\ \mathrm{N2N3N2}\,* \end{vmatrix}; \qquad \mathbf{Am(3)} = \begin{vmatrix} \mathrm{N2N2N3} \\ \mathrm{N1N3N2}\,* \\ \mathrm{N1N2N3} \end{vmatrix}$$

and introduce the twist transformation:

$$\phi\, X_A\, \phi^{-1} = \eta;\ \phi^{-1}\, Y_A\, \phi = \eta^{-1}$$

$$X_A\, \phi^{-2}\, Y_A = \phi^{-2};\ Y_A\, \phi^{+2}\, X_A = \phi^{+2}.$$

Vector η is the wave vector of propagation of twistor conformal field ϕ. Dual field can be transformed in the same way, but in state Av = N1N2N3* that can be verified directly. Field ϕ has a λ-form characteristic.

Table 2.7 Amino acids conformal field in a mixed gauge: Av. A1(N1N2N3) A2, A3_A2′, A3′(N3N1N2)

A.D(**GCA**)A,Q_P,E(**AGC**)	A.G(**GCG**)A,R_P,G(**GGC**)
A.A(**GCC**)A,P_P,A(**CGC**)	A.V(**GCU**)A,L_P,V(**UGC**)
C.S(**UGC**)C,A_G,S(**CUG**)	C.L(**UGU**)C,V_G,F(**UUG**)
D.A(**GAC**)D,T_N,A(**CGA**)	D.V(**GAU**)D,I_N,V(**UGA**)
E.E(**GAA**)E,K_K,E(**AGA**)	E.G(**GAG**)E,R_K,G(**GGA**)
F.S(**UUC**)F,S_F,S(**CUU**)	F.F(**UUU**)F,F_F,F(**UUU**)
G.E(**GGA**)G,D_G,D(**AGG**)	G.G(**GGG**)G,G_G,G(**GGG**)
G.A(**GGC**)G,A_G,A(**CGG**)	G.V(**GGU**)G,V_G,V(**UGG**)
H.P(**CAC**)H,T_N,P(**CCA**)	H.L(**CAU**)H,I_N,L(**UCA**)
I,N(**AUA**)I,Y_L,N(**AAU**)	I.T(**AUC**)I,S_F,T(**CAU**)
I.I(**AUU**)I,F_F,I(**UAU**)	
K.K(**AAA**)K,K_K,K(**AAA**)	K.R(**AAG**)K,R_K,R(**GAA**)
L.H(**CUA**)L,Y_L,H(**ACU**)	L.R(**CUG**)L,C_L,R(**GCU**)
L.P(**CUC**)L,S_F,P(**CCU**)	L.L(**CUU**)L,F_F,L(**UCU**)
L.Y(**UUA**)L,Y_L,Y(**AUU**)	L.C(**UUG**)L,C_L,C(**GUU**)
M.S(**AUG**)M,C_L,S(**GAU**)	
N.T(**AAC**)N,T_N,T(**CAA**)	N.I(**AAU**)N,I_N,I(**UAA**)
P.H(**CCA**)P,Q_P,Q(**ACC**)	P.R(**CCG**)P,R_P,R(**GCC**)
P.P(**CCC**)P,P_P,P(**CCC**)	P.L(**CCU**)P,L_P,L(**UCC**)
Q.Q(**CAA**)Q,K_K,Q(**ACA**)	Q.R(**CAG**)Q,R_K,R(**GCA**)
R.K(**AGA**)R,D_G,N(**AAG**)	R.R(**AGG**)R,G_G,S(**GAG**)
R.Q(**CGA**)R,D_G,H(**ACG**)	R.R(**CGG**)R,G_G,R(**GCG**)
R.P(**CGC**)R,A_G,P(**CCG**)	R.L(**CGU**)R,V_G,L(**UCG**)
S.T(**AGC**)S,A_G,T(**CAG**)	S.M(**AGU**)S,V_G,I(**UAG**)
S.Y(**UCA**)S,Q_P,*(**AUC**)	S.C(**UCG**)S,R_P,W(**GUC**)
S.S(**UCC**)S,P_P,S(**CUC**)	S.F(**UCU**)S,L_P,L(**UUC**)
T.N(**ACA**)T,Q_P,K(**AAC**)	T.S(**ACG**)T,R_P,R(**GAC**)
T.T(**ACC**)T,P_P,T(**CAC**)	T.I(**ACU**)T,L_P,M(**UAC**)
V.D(**GUA**)V,Y_L,D(**AGU**)	V.G(**GUG**)V,C_L,G(**GGU**)
V.A(**GUC**)V,S_F,A(**CGU**)	V.V(**GUU**)V,F_F,V(**UGU**)
W.W(**UGG**)W,G_G,C(**GUG**)	
Y.S(**UAC**)Y,T_N,S(**CUA**)	Y.L(**UAU**)Y,I_N,L(**UUA**)
.(**UAA**)*,K_K,*(**AUA**)	*.*(**UAG**)*,R_K,*(**GUA**)
.(**UGA**)*,D_G,Y(**AUG**)	

Figure 2.5 shows the duplication of field ϕ of its own connection elements. The duplication process is divided into three phases I, II and III; in each of them the charge conservation law has a different sense.

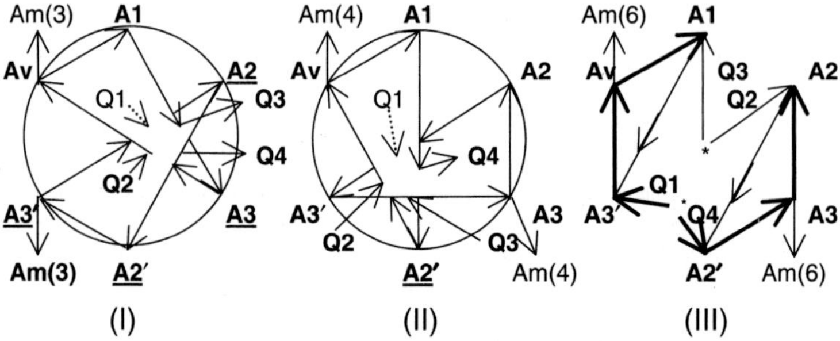

Fig. 2.5 Conformal field duplication. Field of type ω is shown; field of type M in alphabetical interpretation is its copy.

When obtaining formulae of external differentiation: *d* symbol determines the change of external metric of field ϕ; ∂ symbol designates increase of ordinal type; symbol Q = {**Q1, Q2, Q3, Q4**} determines charge of field ϕ; and symbol *m* indicates the influence of marks.

At each phase, charge variation is equal to δQ = 0, and is determined by the sum of its changes.

Phase I. Charges Q are invariants of Lie group: they express independence of basic amino acid quantity from direction of DNA transcription and RNA translation; and form ordinal type ω_1. Incident charges do not exist. The external metric of field ϕ is induced by mixed calibration.

$$dQ = 2^+ 4^- = -2; \mathbf{Q2} + \mathbf{Q4} + \mathbf{Q1}^* = \mathbf{Q3};$$
$$\partial Q = 2 + 4 - 3 - 1 = +2$$

Phase II. Charges Q are incident to straight-line segment and form an ordinal type ω_2. Field ϕ has four components: Am(4) = Av. A1, A2, A3. Increase in the quantity of field components compensates loss of ordinal type ω_1.

$$dQ = 5^+1^- = +4; \mathbf{Q2} + \mathbf{Q3} + \mathbf{Q1}^* = \mathbf{Q4};$$
$$\partial Q = 2 + 3 - 1 - 4 = 0;$$
$$\partial Qm = (\underline{A2'} - A3) = -2; ddQ^- = -4 + 2 = -2.$$

Phase III. There is no external metric. Charges Q are ordinal numbers of ordinal type ω_2. Field ϕ is in a duplication stage.

$$dQ = 0; \partial Qm = (A3' - \underline{A2'}) = +2; ddQ^+ = -2.$$

The quantity of basic amino acids on segment A3_A2′ is less than as in segment A1_Av, therefore, terminator poles move relatively each other. Since **Q2** > **Q1**, poles diverge. These segments determine the Boolean elements:

$$\partial(A1 - Av) = 1, \partial(A3 - A2') = 1' = 0.$$

From here, $\partial\omega_2 = \mathbf{Z}_2$. Complete differential

$$\phi(A2) + \phi(A3) - \phi(A3') = 2\phi(*); \phi(*) = \mathbf{Z}_2$$

is a consequence of gauge invariance of incident charges **Q1**, **Q4**.

Ordinal numbers are:

$$\mathbf{Q1} = Q(A2 + A3 - A3' + *);$$
$$\mathbf{Q2} = Q(A3' - A2 - *);$$
$$\mathbf{Q3} = Q(A1 + A2');$$
$$\mathbf{Q4} = Q(A3 + A2' - Av + *); \mathbf{Q4}(*) = -1.$$

Ordinal type $\omega_3 = \omega_2 + \mathbf{Z}_2$ is accumulating, it does not contain a terminator and determines the linear growth of field ϕ from four to six components:

$$CFAA = \phi = Am(6) = Av.\ A1, A2, A3\_A2', A3'$$

Equality **Q2** = **Q3** determines the scale invariance of field ϕ dynamical transformations. It is clear from Fig. 2.5(III) that 10-parametrical Poincaré group is a group of these transformations.

The duplication process of dual conformal field with metric 3^+3 has a character of twistor auto-oscillations because $\partial\omega_1 = \mathbf{Z}_2$.

Conformal field of amino acids has an evident alphabetical interpretation.

And, finally, due to the fact that **Lorentz group is six-parametrical Lie group and conformal field ϕ has infinite group of transformations, also including Lorentz and Poincaré groups, it is possible to assume that the conformal field ϕ and quantized gravitational field are physically identical.**

Let us prove that the **contracted curvature tensor R_{ik} [18, 19] induces duplication of curvature tensor R_{iksm} in a four-dimensional space.**

Let us represent R_{ik} as:

$$R_{ik} = (\Gamma^s{}_{ik;s} + \Gamma^s{}_{ik}\Gamma^m{}_{sm}) - (\Gamma^s{}_{is;k} + \Gamma^m{}_{is}\Gamma^s{}_{km}) = R_{ik}{}^+ - R_{ik}{}^-.$$

Let us represent Christoffel symbols on matrices of groups GL(2, L):

$$R_{ik}{}^+(GL) = \begin{vmatrix} s & s \\ i & k \end{vmatrix} + \begin{vmatrix} s & 0 \\ i & k \end{vmatrix} \begin{vmatrix} m & 0 \\ s & m \end{vmatrix};$$

$$R_{ik}{}^-(GL) = \begin{vmatrix} s & k \\ i & s \end{vmatrix} + \begin{vmatrix} m & 0 \\ i & s \end{vmatrix} \begin{vmatrix} s & 0 \\ k & m \end{vmatrix}.$$

(There is no summation by **s** index here.)

After multiplication and summation of matrices, let us represent $R_{ik}^{\pm}$ (GL) as:

$$R_{ik}^{+}(GL) = \begin{vmatrix} \alpha & d \\ b & c \end{vmatrix}; \qquad R_{ik}^{-}(GL) = \begin{vmatrix} \alpha & g \\ e & f \end{vmatrix},$$

where **b** = **im** + **ks** + **i**,...

Let us rearrange the matrix columns by involutive element * and take into account that differentiation by index 'α' can be omitted; hence, we obtain reflection:

$$R_{ik}^{+} \rightarrow {}^*\Gamma^{d}{}_{cb};\ R_{ik} \rightarrow {}^*\Gamma^{g}{}_{fe}.$$

Let us identify 20 independent components of curvature tensor R_{iksm} with 20 observed amino acids, involutive element * is identified with a terminator. We obtain a basis of finite projected plane P(4). Identity ${}^*\Gamma^{d}{}_{cb} = {}^*\Gamma^{g}{}_{fe}$ in P(4) is implemented when two free parameters are present; it induces identity $R_{ik}^{+} = R_{ik}^{-}$ of bundle R_{ik} and duplication of curvature tensor $R_{iksm} = R_{smik}$, which was to be proved.

Let us select 20 triangles of connection by the group S(3) as a basis of finite projected plane P(4):

Am(N1N2N3), Am(N2N3N1), Am(N3N1N2)

with a condition that (*) = NNN. Let us transfer the affine connection

(122, 233, 311), (121, 232, 313), (113, 221, 332)

into projected one

(122, 222, 211), (121, 212, 111), (112, 111, 221).

As the next step, we divide 35 of 36 triple points of bundle (*):

Am(N1N2N2), Am(N2N2N2), Am(N2N1N1)
Am(N1N2N1), Am(N2N1N2), Am(N1N1N1)
Am(N1N1N2), Am(N1N1N1), Am(N2N2N1)

by 105 points and place them on coordinate plane. Let us determine one triple point G(**GGC**), G(**GGG**), P(**CCG**) with $Q_4 = -21$ as a pole of Cartesian coordinate system. We obtain finite projected plane of the order four with factor-metric $6^{+}4^{-}Q_4(M, -D)$ = 10-parametrical Poincaré group.

Tables 2.8 and 2.9 represent basis P(4) of 36 triple points of bundle (*).

Let us select the mobile, synchronous Lorentz-invariant basis of finite projected plane P(2) in built basis:

$$Q_4(-* + S + V) = M(41); Q_4(-P\ \ H + T) = -M(41);$$
$$Q_4\{(+S + L + F) + (+S - R + V) + (-R + G + A)\} = K(41);$$
$$Q_4\{(+W + G + V) + (+S + A - Q) + (-N + T\ Q)\} = K(41).$$

Four particles (W, G, V), (S, A, Q), (S, R, V), (P, H, T) with $Q_4 = 32^{**}$ form the ordinal type ω_1 of RNA-polymerase enzyme. Field M(41) is σ-factor. Two fields K(41) correspond to two ribosome sub-particles of ordinal type ω_2. Matrix 'α' is dual to plane P(4) in field **GF**(167).

Table 2.8 Twenty triple points on the basis of finite projected plane P(4)

*(**UAA**) → N(**AAU**) ↖ ↙ I(**AUA**)	S(**UCU**) → L(**CUU**) ↖ ↙ F(**UUC**)	*(**UAG**) → S(**AGU**) ↖ ↙ V(**GUA**)
S(**UCC**) → P(**CCU**) ↖ ↙ L(**CUC**)	*(**UGA**) → D(**GAU**) ↖ ↙ M(**AUG**)	S(**UCA**) → H(**CAU**) ↖ ↙ I(**AUC**)
P(**CCA**) → H(**CAC**) ↖ ↙ T(**ACC**)	S(**UCG**) → R(**CGU**) ↖ ↙ V(**GUC**)	N(**AAC**) → T(**ACA**) ↖ ↙ Q(**CAA**)
S(**AGC**) → A(**GCA**) ↖ ↙ Q(**CAG**)	R(**CGC**) → A(**GCC**) ↖ ↙ P(**CCG**)	R(**CGA**) → D(**GAC**) ↖ ↙ T(**ACG**)
C(**UGC**) → A(**GCU**) ↖ ↙ L(**CUG**)	R(**CGG**) → G(**GGC**) ↖ ↙ A(**GCG**)	R(**AGA**) → E(**GAA**) ↖ ↙ K(**AAG**)
Y(**UAU**) → I(**AUU**) ↖ ↙ L(**UUA**)	R(**AGG**) → G(**GGA**) ↖ ↙ E(**GAG**)	Y(**UAC**) → T(**ACU**) ↖ ↙ L(**CUA**)
W(**UGG**) → G(**GGU**) ↖ ↙ V(**GUG**)	V(**GUU**) → L(**UUG**) ↖ ↙ C(**UGU**)	

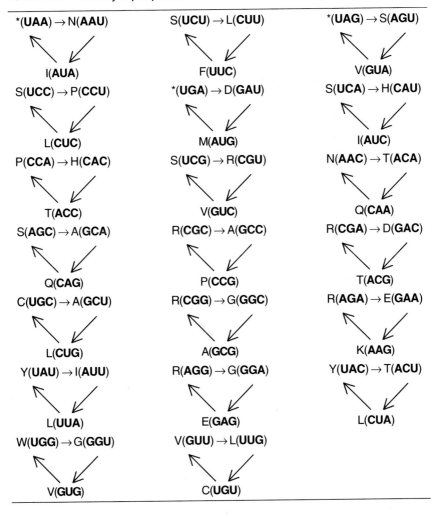

Table 2.9 Thirty-six triple points of terminator bundle in a finite projected plane P(4)

I(**AUU**),F(**UUU**),*(**UAA**)	R(**AGG**),G(**GGG**),E(**GAA**)
I(**AUA**),Y(**UAU**),K(**AAA**)	R(**AGA**),E(**GAG**),K(**AAA**)
N(**AAU**),K(**AAA**),L(**UUA**)	K(**AAG**),K(**AAA**),G(**GGA**)
T(**ACC**),P(**CCC**),Q(**CAA**)	*(**UAA**),K(**AAA**),I(**AUU**)
T(**ACA**),H(**CAC**),K(**AAA**)	Y(**UAU**),I(**AUA**),F(**UUU**)
N(**AAC**),K(**AAA**),P(**CCA**)	L(**UUA**),F(**UUU**),N(**AAU**)
W(**UGG**),G(**GGG**),V(**GUU**)	S(**UCC**),P(**CCC**),L(**CUU**)
C(**UGU**),V(**GUG**),F(**UUU**)	S(**UCU**),L(**CUC**),F(**UUU**)
L(**UUG**),F(**UUU**),G(**GGU**)	F(**UUC**),F(**UUU**),P(**CCU**)
E(**GAA**),K(**AAA**),R(**AGG**)	V(**GUU**),F(**UUU**),W(**UGG**)
E(**GAG**),R(**AGA**),G(**GGG**)	V(**GUG**),C(**UGU**),G(**GGG**)
G(**GGA**),G(**GGG**),K(**AAG**)	G(**GGU**),G(**GGG**),L(**UUG**)
A(**GCC**),P(**CCC**),R(**CGG**)	Q(**CAA**),K(**AAA**),T(**ACC**)
A(**GCG**),R(**CGC**),G(**GGG**)	H(**CAC**),T(**ACA**),P(**CCC**)
G(**GGC**),G(**GGG**),P(**CCG**)	P(**CCA**),P(**CCC**),N(**AAC**)
L(**CUU**),F(**UUU**),S(**UCC**)	R(**CGG**),G(**GGG**),A(**GCC**)
L(**CUC**),S(**UCU**),P(**CCC**)	R(**CGC**),A(**GCG**),P(**CCC**)
P(**CCU**),P(**CCC**),F(**UUC**)	P(**CCG**),P(**CCC**),G(**GGC**)

Let us examine the CFAA propagation in Lorentz gauge:

$$\phi - * = YM_4(Am)^+ + YM_4(Am)^- + M.$$

We will examine field ϕ as temporal, field M as dilatation and Yang-Mills fields $YM_4(Am)^+$, $YM_4(Am)^-$ as transversal CFAA parts.

During DNA transcription, field ϕ has only one component M, which is σ-factor. Then the anticommutator $\{YM_4(Am)^+, YM_4(Am)^-\} = 1$. Fields $YM_4(Am)^+$ and $YM_4(Am)^-$ are (+) and (–) DNA strands. The dilatation part of CFAA is the RNA strand of field M, which is homeomorphous to (+) DNA strand.

During RNA translation, field ϕ has six components and complete number of states equal to 64, because $\phi(*) = \mathbf{Z}_2$. Field M has only one component. Basis Am* is equal to 21 and determines basis P(4). Because basis (ϕ–*) is equal to seven and determines basis P(2) then anticommutating fields $YM_4(Am)^+$, $YM_4(Am)^-$ can depend only on nucleotide triplet as coordinates in CFAA charge space. Matrix 'α' puts a connection on Yang-Mills fields:

$$\{YM_4(Am)^+, YM_4(Am)^-\} = YM_4(Am)^{'\alpha'}$$

and is the representation of genetic code in self-dual basis of amino acids. It is possible to restore the genetic code simply by reverse reflection K $\rightarrow$ **A**, F $\rightarrow$ **U**, P $\rightarrow$ **C**, G $\rightarrow$ **G** using matrix 'α'.

In Hamiltonian gauge $\phi = 0$. Then matrix 'α' is metric tensor g_{ik} of CFAA. Basis of CFAA contains 20 amino acids, three terminators and a modified methyonine M* = –M–*, which determines the start of RNA translation; 24 components in total. Basis of metric tensor g_{ik}, expressed through the sequence of nucleotide triplets in its own charge space, contains 72 nucleotides and is identified with tRNA basic nucleotide sequence. In tRNA structure, it is possible to select paired and unpaired nucleotide strands, which are identified with derivatives of metric tensor.

Metric tensor of six-dimensional Riemann space with a signature $4^{+}2^{-}$ has $21 \times 4 \times 2 = 168$ components and determines tRNA bundle in four-dimensional space. In six-dimesional space, tensor R_{iksm} has 105 components of metric tensor g_{ik} of plane P(4). Then

$$Q(g_{ik})_{P(4)} + Q(g_{ik})_{'\alpha'} = Q(g_{ik})_{\phi} = Q(\text{tRNA}).$$

Molecules of tRNA are the components of CFAA metric tensor and perform the functions of connection, i.e., they transfer the information of RNA matrix upon the protein along the trajectories of plane P(4), in which RNA is located. Genetic code is a CFAA metric tensor in a proper charge space.

This fact explains the high stability and universality of the genetic code. The availability of set of gauge-equivalent tRNA trajectories completely removes the problem of genetic code degeneration, because quantity of information contained in RNA is identical to the quantity of information contained in the movement group of isoaccepting tRNA set and is equal to the sum of quantity of information transmitted to protein and residual information located in the tangent plane P(4) of ribosome movement.

When changing gauge, six components of fields φ reduce themselves; the field **GF**(3) induces an external conformal field containing only charges ±G of Hadamard matrix. The external conformal field is determined as CFAA coloured space; it is identical to CFAA and generates structures on the base of DNA and RNA molecules. CFAA basis consists of antiproton –G, i.e., bound proton and nucleotides **A, G, C, T, U**.

Conformal field of amino acids, identical to quantized gravitational field, can be accepted as a general theoretic basis for the explanation of biological processes. DNA transcriptions and RNA translations are determined as processes of CFAA propagation in its own charge space. Cell division is one of the CFAA auto-radiation forms.

Informational Structure of DNA, RNA and Protein

3

Chapter

Irreversible transformations of set

Let us examine a set of elements Ω with coincided elements. We will consider coincided elements as marked elements and the unmarked elements can divide all the marked ones. If all elements of a set are identical then they form a bundle base. **Bundle over the base represents information.**

A set of orthogonal vectors is the simplest bundle. Any bundle is built with the aid of generating operator. The authors see the generalization of quantum mechanics in this fact, where operator of energy of a physical system is a generating operator.

For example, the set of identical elements of order 16 m^2 is transformed in instanton of order 64 m^3 with the aid of an operator $Y_{S/S}$ and with the sum of elements16 m^2. Initial set, at first, is transformed into Hadamard matrix of order 4 m **spontaneously**, because each line of Hadamard matrix is a proper vector of homogenous field of units as a set of trivial characters of identical elements. Next, instanton is built on Hadamard matrix.

Thus, the first ordering occurs spontaneously; and the following ones – with the aid of generating operator.

Ordered set of elements always has a certain holonomy connection group. It comes from the fact that the ordering process is impossible without simultaneous transfer of several elements of an initial set. Transfer of two elements (marked and unmarked ones) is the simplest case. Numerated set is that of marked elements which are separated by unmarked ones. Holonomy connection group is built on marked as well as unmarked elements. Each element of holonomy connection group forms a certain ordinal type, which consists of a fundamental group of subset elements of the initial set elements. An ensemble of fundamental groups formed during this process is bound by holonomy connection group of initial set; holonomy connection groups themselves are bound by an instanton. All elements of the set are different in their ordinal types, but they transform synchronously to the other elements of the set. Thus, the presence of identical elements in a set gives an opportunity to introduce synchronous transformations.

Equations of classical mechanics are written for synchronous transformations of physical values (time as a parameter of transformation). Lagrange function and its derivatives are generating operators. Initial physical values are taken as base values, change of the base in time is bundle; initial base also becomes a parameter of movement and there is no real bundle.

The behaviour of dynamic system in gauge-invariant models is also determined by the initial data. During movement process, new dynamic variables are formed spontaneously, as the dimension of the base increases. Multitude of elements of new base relatively to multitude of elements of initial base forms a true bundle, which is irreversible and works as a memory.

All this occurs during instanton construction. Unmarked elements of set are new dynamic variables. Sometimes these elements are called fictitious points, gaps, quasi-particles, virtual particles, etc. Taking into account the fact that the theory of latent parameters is of little interest, moreover, shielding (including that of dynamic variables) is not possible in gravitational field; it is necessary to use non-linear operators, which are always irreversible, for construction of generating operator.

Self-acting of these operators is revealed by the fact that the initial dynamic variables are united into subsets and only then are new dynamic variables, and their quantity is greater; new additional bonds appear among them and, consequently, a holonomy connection group. Nevertheless, 'gaps' remain (often as a factor set) but they are full dynamic variables.

Informational structure of DNA

Ordering by involvement is possible only in a physical Fermi systems. Physical elements with opposite spins form a bundle base.

Conformal field of amino acids with a spin of 5/2 has six states in a six-dimensional physical space: **four** informational states, **one** factor state and **one** field **GF**(3*) of information parallel transfer equivalent to pairing of informational symbols. We note that the classical gravitational field has a spin equal to two.

Gauge equivalency of charges **Q1** and **Q4** is revealed not only in the fact that it removes degeneracy of genetic code but also allows introducing of a space with marked point. The state of base **Q1** is marked by charges **Q4**; the state of base **Q2** separates the state of base **Q1**; and the state of base **Q3** is a gap. Charges **Q1, Q2, Q3, Q4** are ordinal numbers and have correspondingly 1, 2, 3 and 4 informational states.

Figure 3.1 shows the structure of a DNA basic strand. DNA backbone is formed by alteration of two basic links: phosphate basis of nucleotide C without marked point and furanose basis of nucleotide U with marked point CH_2. Basic nucleotide C is symmetric and does not have an orientation along the basic strand while basic nucleotide U is significantly oriented.

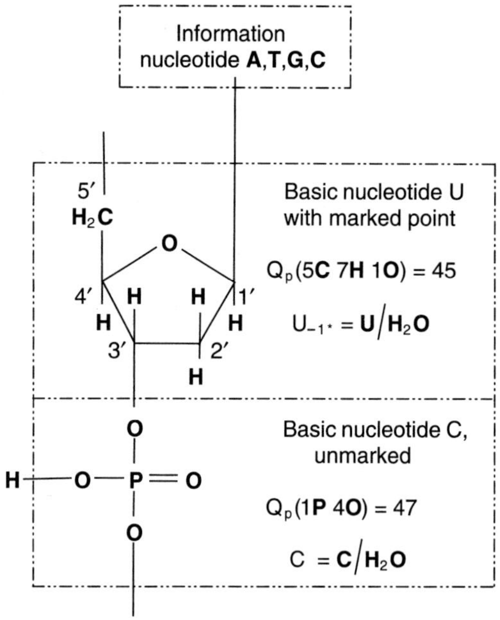

Fig. 3.1 Structure of a DNA bundle basis. Basic nucleotide U and basic nucleotide C, which are dual, form a base of DNA bundle. Basic nucleotide C is not marked as a nucleotide but contains primary radical **H**, which is a unit of inversion of fine structure constant relatively divisor UCU. Basic nucleotide U has a complex marked point – informational nucleotide, which is a secondary DNA radical. Nucleotide U forms a right ideal of DNA bundle relatively to nucleotide C. This ideal contains marked point 5' (**H_2C**) of DNA methylation leaving bundle centre (nucleotide C) on the left. Orientation of basic strand 5' 3' of right ideal U leaves **two degrees of freedom** 1' and 2' by linking of informational nucleotides to furanose ring. Therefore, two fields **GF**(137) of the bundle of complementary pairs of four informational nucleotides relatively marked 1' are required. Factor metric of basic strand is equal to $2^+ 8^-$ (**H_2O**); Q_p(**H_2O**) = 10.

Each mark of the furanose ring orientation expresses identity of the ordinal numbers: only **one** informational nucleotide (**Q1** = **Q4**) is attached to 1' mark; marked point $\mathbf{CH_2}$ (**Q2** = **Q3**) **is not attached** to 2' mark; only basic nucleotide C **dual** to basic nucleotide U (**Q3** = **Q2**) is attached to 3' mark; marked point $\mathbf{CH_2}$ (**Q2** = **Q3**) is attached to 4' mark; a copy of the unmarked basic nucleotide C is attached to 5' $\mathbf{CH_2}$.

Complementary DNA pairs are shown in Fig. 3.2.

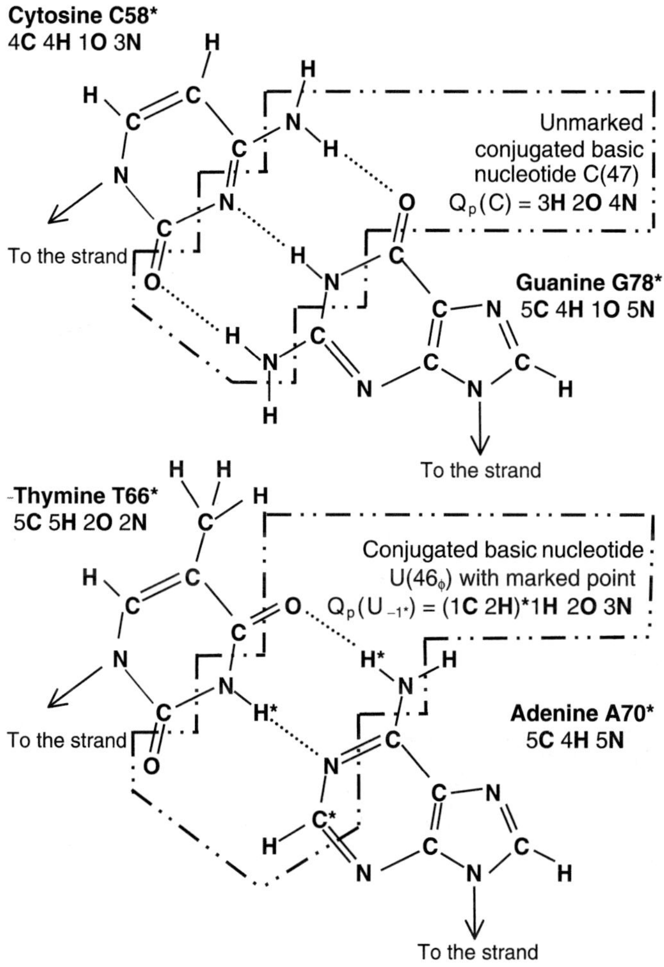

Fig. 3.2 Complementary DNA pairs. Pair **CG** is formed during the parallel transfer of basic exchange charge C(47) by field **GF**(3). Pair **AT** is formed during the parallel transfer of marked point $\mathbf{CH_2}$ of basic exchange nucleotide U_{-1}* by field **GF**(5) = $\mathbf{CH_2}$/**GF**(3).

When pairing, informational nucleotides form a field **GF**(137) of inverse fine structure constant:

$$\mathbf{C}58^* + \mathbf{G}78^* + \mathbf{GF}(3) = \mathbf{T}66^* + \mathbf{A}70^* + \mathbf{GF}(3)$$
$$= \mathbf{T}66 + \mathbf{A}70 + \mathbf{GF}(5^{**}) = \mathbf{GF}(137).$$

Let us calculate the differential four-index commutator of representation of fields **GF**(137) using chemical formulae of DNA informational nucleotides:

$$\Delta_{\mathbf{CGAT}} = [\mathbf{C} + \mathbf{G}]_{\mathrm{DNA}} - [\mathbf{A} + \mathbf{T}]_{\mathrm{DNA}} = 1\mathbf{N} - 1\mathbf{C} - 1\mathbf{H} = \mathbf{n},$$

where $\mathbf{n}$ is neutron. Since $Q_p(\mathbf{n}) = \mathbf{Z}_2$ and from here $Q_p(\mathbf{n}) = \phi(*)$, then connection $\Delta_{\mathbf{CGAT}}$ is equivalent to neutral exchanging current between different complementary pairs.

Figure 3.3 shows the DNA bundle over base UC. Complementary symbols are turned at $\pi/2$ angle and written in a mixed gauge:

$$\{\mathbf{A}, \mathbf{T}^{*\pi}\}, \{\mathbf{T}, \mathbf{A}^{*\pi}\}, \{\mathbf{C}, \mathbf{G}^{*\pi}\}, \{\mathbf{G}, \mathbf{C}^{*\pi}\}.$$

Base symbols $U^{\pi} = \cap$, $C^{\pi} = \supset$.

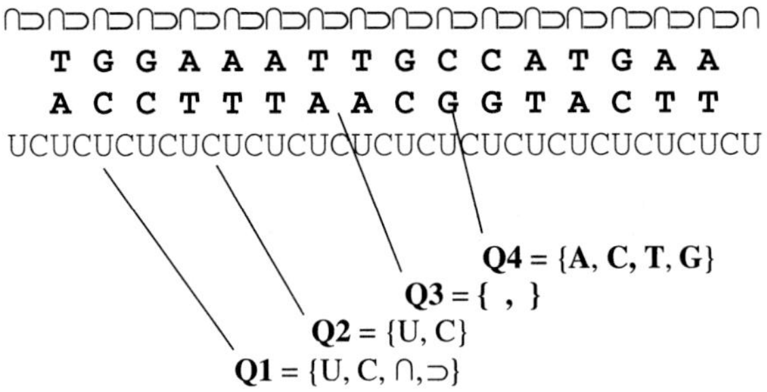

Fig. 3.3 Informational structure of DNA

Supercompression of DNA

The presence of a factor set of the gap in an ordinal type ω_3 gives an opportunity to compress the DNA space. This compression can be divided into two phases:

1. Shift of complementary strands with filling of gaps.
2. Mutual enclosure of complementary strands.

DNA compression by the right shifting of one complementary strand turns DNA into complex, dynamic DNA. The obtained linear strand of nucleotides are interpreted as a transversal-triplet RNA; as well as linear, duplicated DNA.

Metric $4^+ 2^-$ of amino acid conformal field limits the choice of mutual enclosure of the DNA strands. This enclosure induces right twist of informational nucleotides as shown on Fig. 3.4. Correspondingly, left twist of informational symbols leads to linear, duplicated DNA ordered in inverse direction. Invariance of metric $4^+ 2^-$ preserves the DNA integrity during compression.

ATCGCGTATATAATATCGGCGCTAATCGTATA

UCUCUCUCUCUCUCUCUCUCUCUCUCUCUCUCU

Fig. 3.4 DNA shift and compression

Let us write the obtained transversal triplets' nucleotides accounting for the nucleotide complementarity (nucleotide U is self-conjugated):

UTC* = UTG	CT*U* = **CAU**
UGC* = **UGG**	CG*U* = **CCU**
UCC* = **UCG**	CC*U* = **CGU**
UAC* = **UAG**	CA*U* = CTU

Triplets UTG and CTU correspond to virtual Ξ-amino acids but can also be observed. Conformal field of amino acids has duplicated and, in accordance with the genetic code, looks as shown below:

Ξ.Ξ,Ξ,Ξ_G,Ξ	H.L,I,S_F,T
W.W,G,V_G,V	P.L,L,S_F,P
S.C,R,V_G,L	R.L,V,S_F,A
.,S,V_G,I	Ξ.Ξ,Ξ,Ξ_F,Ξ

Each of the two Ξ-amino acids are distinct in position A2'(N3N3N1) and are G, F analogues. CFAA is compressed to two amino acids with Q_p(G, F) = 50. This quantity of Galois fields is contained by field **GF**(167), therefore, DNA compression by 50 times is caused by transition **GF**(137) → **GF**(167).

Linear, duplicated DNA is a protein consisting of two amino acids G, F and, therefore, is compacted. Gauge invariance of charges **Q1** and **Q4** gives reason to assume that DNA rolling cannot be similar to statistical rolling up into a ball. In the state A2', CFAA is in two possible states, therefore, DNA rolling is a trigger transition with finite time of the life of rolled state.

Unlike classic gravitational field, for which collapse, i.e., field compression, is irreversible because of the uniquity of this state, CFAA is reversible because it has two divergence poles G and F.

Linear, duplicated DNA can be read and transferred into observed RNA with partial cut of introns, the presence of which is connected with duplication of the informational sequence. This gives new proteins and possibility for the evolutional development of a biological system.

Identity **Q1** = **Q4** limits the time of DNA residing in its compressed state, whereas

possible identity **Q1** = **Q3**(?)

and parallel transferred identity **Q2** = **Q4**(?)

in totality with a norm **Q2** = **Q3**

form poles **Q1** and **Q2** of DNA duplication on the basis of identity Q_p(**U**) = Q_p(**C**), and DNA residing time reduces by twice, however, total DNA duplication time remains invariable (it is necessary to have additional time for returning into initial state **Q1** = **Q4**). Let us note that the identity **Q1** = **Q3**(?) is an obstacle for subsequent DNA compression and if transversal-triplet RNA can be read by biological system (**three-dimensional RNA in one-dimensional DNA position is read**), then the obtained protein will stimulate the subsequent DNA compression in state **Q1** = **Q4**.

If linear, duplicated DNA can be read, then the identity **Q1** = **Q4** will not be broken (it will be read as that one-dimensional DNA strand consists of four nucleotides). It is possible in this case to obtain an absolutely new protein as a feature of the evolutionary development of the biological system.

CFAA being in state G, F compensates the DNA transition in still more compressed state, which can be called quasi-scalar DNA. Basic nucleotides U, C become superstructure over former informational nucleotides **A**, **T**, **G** and **C**. It occurs because quasi-scalar DNA (Fig. 3.5) can be only in two states.

T	G	G	A	A	A	T	T	G	C	C	A	T	G	A	A
A	C	C	T	T	T	A	A	C	G	G	T	A	C	T	T

Fig. 3.5 Quasi-scalar DNA in two possible states

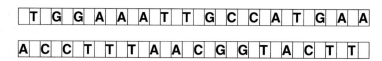

Let us prove that the **quasi-scalar DNA is a conformal field of amino acids with poles of divergence K, R and Q4(K) – Q4(–R) = 96**.

During the conversion of DNA base, conformal field comes into dual state. Instead of four basic nucleotides, only one **mononucleotide A** remains, **group of generating operator G96** and structure of cell metaphase spindle with space congruence of CFAA state are formed. It is precisely that this fact was to be proved.

Lets us remind you that the Schwarzschild field is compressed into a sphere of radius r_g, i.e., it has only one compressing pole. Compressed CFAA has two compressing poles – let us find them.

Dual CFAA in position A1'(N1N2N2) has six identical states of amino acid R, which form a new bundle base. Marked point R.R = GG is a superstructure over the former base G. Using Table 2.1 of the genetic code, it is necessary to find the predecessors of marked point R.R, which form the bundle over RR. These are amino acids K(**AAG**) and P(**CCG**). Amino acid P contains its own marked point* and is a bundle over K, consequently, triplet **AAA** is a basis of amino acid K with marked point **AA**. Mononucleotide **A** is a base of dual CFAA bundle as a factor set during CFAA duplication. DNA turns into RNA via the conversion of CFAA base:

1. Basic nucleotide U(DNA) turns into basic nucleotide T(RNA): $U_{-1*} \rightarrow T55_{-2}$.
2. Informational nucleotide **T**(DNA) turns into informational nucleotide **U**(RNA).

This transition is caused by a shift on 8* in base U (Fig. 3.6).

⊥⊃⊥⊃⊥⊃⊥⊃⊥⊃⊥⊃⊥⊃⊥⊃⊥⊃⊥⊃⊥⊃⊥⊃⊥⊃⊥⊃⊥⊃⊥⊃⊥⊃⊥

U G G A A A U U G C C A U G A A

Fig. 3.6 Structure of informational RNA. Base symbols $T^{\pi} = \perp$, $C^{\pi} = \supset$.

Involutive transition in CFAA base **T**(U) ↔ **U**(T) is normal CFAA divisor and it determines metric $2^+ 2^-$ of RNA charge space in informational RNA basis: pair **A**, **U** is self-conjugated; pair **G**, **C** remains complementary-conjugated and transition **G** ↔ **C** does not change dynamic informational protein capacity, and this fact was mentioned in Chapter 1. The volume of group of protein synthesis generating operator is equal to:

$$\mathbf{Q4}(K) - \mathbf{Q4}(-R) = 96$$

and is a normal divisor of inverse CFAA base equal to 64***GF**(3) = 192**A** (three bases of CFAA: initial, dual and inverse). Group **G96** has own involutive normal divisor, therefore, identity **Q1 = Q4** is interpreted as **one** group **G**96 of protein synthesis generating operator with RNA basic congruence nucleus 32**A** and with metric $2^+ 2^-$ (two involutive normal divisors of inverse base 192**A**) of **four-dimensional** RNA charge space. Group **G**96 has cyclic indexes 1, 2, 3, 4, 6; and therefore, is a suitable group of protein synthesis generating operator. Let us emphasize that group **G**96 is formed before the moment of DNA compression finish.

Group **G**96 performs additional compression of linear, duplicated DNA by more than 96 times. Quasi-scalar DNA is the maximum rolling of initial DNA by 50×96 times.

Let us write a connection identity in the bundle of marked point RR by taking into account the chemical structure of amino acid radical:

Q4(–P*) – **Q4**(–R) = Q_p(1**C** 5**H*** 3**N**) = 32^{-*} – congruence module of generating group **G**96

Q4(K) – **Q4**(–P*) = Q_p(7**C** 15**H*** 1**N**) = 64* – congruence module of CFAA with marked point of **H*** antiproton.

Chemical composition of CFAA congruence module can be obtained by a linear combination of chemical elements of the congruence module of group **G**96 accounting for commutator of 5**H***(3**N**) = 15*-parametrical Lie group of icosahedron. It is clear from here that the structure of a cell metaphase spindle is homeomorphous to icosahedron group and is formed as a result of space congruence of CFAA states that was affirmed.

The point of compressed DNA return into initial state of complementary strands with DNA duplication is the moment of formation of **amino acid conformal field generating operator in an impulse space**. CFAA generating operator is metric tensor of 10-parametrical Poincaré group. Since finite projected planes of orders six and 10 are absent, then this tensor has two poles $2^+ 8^-$ (**H_2O**) and $6^+ 4^-$ (**CH_4**). First pole is connected with DNA and RNA production; the second one is connected with protein production (the polypeptide strand structure is examined below in detail). Then CFAA factor metric is equal to $13^+ 7^- = (6^+ 4^- / 2^+ 8^-)$. Both poles correspond to identities:

$$\mathbf{Q1} = \mathbf{Q4} \to 2^+ 8^- (\mathbf{H_2O}) \{\text{DNA, RNA}\}$$
$$\mathbf{Q2} = \mathbf{Q3} \to 6^+ 4^- (\mathbf{CH_4}) \{\text{PROTEIN}\}.$$

Metric tensor of CFAA generating operator, which we designate as M_ϕ, has 100 components and is a bundle over symmetric tensor of 10-dimensional Riemann space, which contains 55 components. Tensor M_ϕ is formed as a tangent bundle over point RR. The quantity of components of tensor M_ϕ can be expressed through CFAA operators, which must satisfy to principle of relativity. According to Table 2.2 of amino acid charge space, we will find

$$Q_p(M_\phi) = 100 = Q_p(H) + Q_p(Y) = Q_p(S) + \mathbf{GF}(83).$$

It follows from this that the entropies covering field **GF**(83) = $Q_p(H) + Q_p(Y) - Q_p(S)$ of genetic code has a connection in bundle of CFAA generating operator with basic generators H, Y. Generator H is recessive operator, imprecise; whereas generator Y is dominant operator, precise. Operator H of four-dimensional space projection on finite plane in a six-dimensional space with metric 4^+2^- is pseudo-inverse to operator Y of primary instanton generation by matrix of tensor M_ϕ. And however, restoration of tensor M_ϕ on finite time interval occurs with errors forming marked points, which are compensated by CFAA, global irreversibility of operator Y is nevertheless connected with the absence of finite projected plane of the order six.

Operator H(43) = $6^2 + 6 + 1$ is a field operator; however, its inaccuracy in projection of CFAA states on the base of the sixth order is caused by the fact that due to the identity of ordinal numbers **Q2** = **Q3** operator H(43) is realized as a finite projected plane operator H(13) = $6 + 6 + 1 = 3^2 + 3 + 1$ of **field GF**(3) **of information parallel transfer**; therefore, there are two states shifted by 30* **GF**(3) charges.

Operator H(43) induces DNA duplication. In order to prove this, let us examine the autooscillations of transversal-triplet RNA, which are formed during DNA compression with the shift of informational nucleotides to the left and to the right of the basic nucleotide U. Field ϕ is built on **nucleotide triplets** as on **duplication ideals** of DNA:

L(CUC) – basis of DNA duplication (bundle);

S(UCU) – dual basis of DNA bundle;

Y(**UAU**), I(**AUA**), A(**GCG**), R(**CGC**) – DNA informational basis.

The Lorentz-interval equivalency to the connection is a direct cause of DNA duplication:

$$Y + I + A - R = L + S - \phi = H(43)^{-*}.$$

Pair L, I is marked point of operator H(43).

Field $H(43)^{-*}$ induces:

a. duplication of informational nucleotides $2 \cdot I = R + S - \phi =$ **T**66:

$$\begin{aligned} Q_p(\mathbf{A}) &= 70; Q_p(\mathbf{G}) = 78; Q_p(\mathbf{T}) = 66; Q_p(\mathbf{C}) = 58; \\ [\mathbf{A} + 2^+] &= \mathbf{G} - \phi = [\mathbf{G} + 2^+] - 8^+; \\ [\mathbf{T} - 2^+] &= \mathbf{C} + \phi = [\mathbf{C} - 2^+] + 8^+. \end{aligned}$$

Complementary nucleotides obtained an increment of charge ±2, which are **copies** of the initial informational nucleotides.

b. duplication of basic nucleotides 2L = Y + A.

Nucleotides **U**, **C** are marked points of Y operator:

$$Q_p(\mathbf{U}^*) = Q_p(\mathbf{C}^*) = Y(57).$$

In a six-dimensional space with metric 4^+2^-, duplication of informational nucleotides induces duplication of ideal U_{-2} with a parallel transfer of the genetic code:

$$2[C - 4^+] = 2U_{-2} = 2H(43) = \mathbf{GF}(83) + \mathbf{GF}(3).$$

Basic nucleotides $\{U, C, \cap, \supset\}$ are copies of the informational nucleotides $\{\mathbf{A}, \mathbf{C}, \mathbf{T}, \mathbf{G}\}$.

Y and U_{-1}* operators are dual:

$$[C + 4^+] = [(U_{-2} + 2^+) + \phi] = U_{-1}{}^* + \phi = F(UUU) - \phi;$$
$$[C - 4^+] = [(C + 2^+) - \phi] = Y - \phi.$$

Basic nucleotide U_{-1}* is a copy of the ideal U_{-2}, of an **improper marked point** of operator Y. Basic nucleotide C is a dual copy of the ideal U_{-2}, **of a proper marked point** of operator Y. Thus, duplication of the ideal U_{-2} leads to the duplication of basic nucleotides U_{-1}*, C.

Amino acid I is coded by three triplets and amino acid L is coded by six triplets of the genetic code. DNA informational basis leans on half of the bundle basis only; therefore, there is a lack of coding triplets of amino acid I for duplication of DNA informational nucleotides. Therefore, operator H(43) is an imprecise operator and requires a supplement. Operator Y(57) is supplement of operator H(43).

Operator Y(57) **is not a field operator but a connection** in bundle of genetic code amino acids. During DNA replication, operator Y(57) acts recurrently on DNA basic nucleotides and becomes the field operator $[Y + 2^+] = \mathbf{GF}(59)$: $2 \cdot I = [Y + 2^+] + P(2)$. Projection of additional informational nucleotides in backward strand of replicating fork by operator $[Y + 2^+]$ leaves ruptures of factor plane P(2) in DNA. These ruptures which are sewed up by DNA-ligase enzyme. The action of DNA-ligase comes to the restoration of connection Y(57) by the way of recurrent projection of informational nucleotides in plane P(2), because $A = [P(2) + 2^+]$.

Informational structure of protein

Figure 3.7 shows a polypeptide strand structure. The strand's backbone forms a covariant basis of protein bundle: amino acid radicals are located asymmetrically to each other as well as components of basis Q(29).

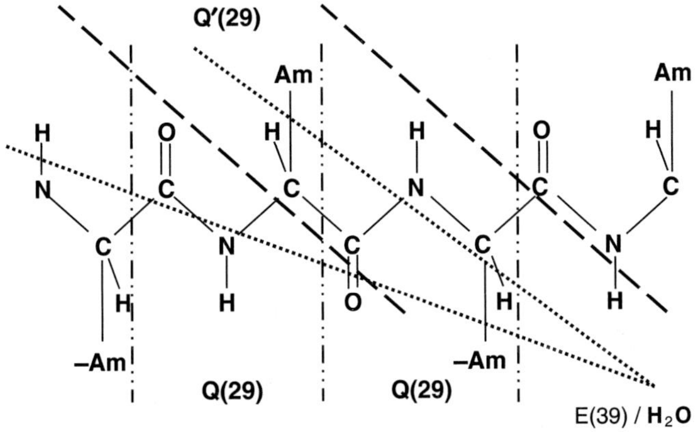

Fig. 3.7 A polypeptide strand structure. Amino acid radicals are designated as Am and –Am. Strand skeleton is formed by basic amino acid Q(29) = Q(39) / $\mathbf{CH_4}$ = **GF**(29) = 2**C** 2**H** 1**N** 1**O**. Virtual bases E(39) / $\mathbf{H_2O}$ = 3**C** 3**H** 1**O** and Q′(29) = Q(39) / $\mathbf{H_2O}$ = 3**C** 4**H** 1**N** are the improper basis of protein bundle. Amino acid radicals are linked up to **C**-inputs of the trigger cell Q(29); radical G(1) = **H** is linked up to **N**-inputs. Trigger Q′(29) has **N**-input. Basis E(39) / $\mathbf{H_2O}$ is diverging. Factor metric of the strand is $6^+ 4^-$ ($\mathbf{CH_4}$).

However, factor metric of a polypeptide strand does not correspond to the commutator of strand $\mathbf{H_2O}$. It means that amino acids obey Fermi statistics and protein is determined as an ordered amino acid sequence with factor-anticommutator $\mathbf{CH_4}$. It is necessary to note that the amino acid sequence in a polypeptide strand has an external (sign–alternative) metric $1^+ 1^-$ as well as an internal metric determined by charge Q_4(Am). Relationship of these metrics determines the tangent protein space, which DNA and RNA are, because protein, in addition to its own basis, has two virtual bundle bases.

Methylation of DNA

RNA being a product of DNA transcription accepts genetic information contained in the DNA. Acceptation act occurs by fixing marked point

O ~ CH_2 in RNA base in position 2' of furanose ring and by cutting introns from the primary transcript.

Let us examine the transformation of RNA into DNA and inverse process by introduction of marked point* **$CH_2 \to O$**. To do it, we divide RNA into neighbouring nucleotide pairs and transform them into complementary pairs of DNA using marked point **CH_2** of DNA methylation. Let us select gauge transformation of nucleotide pairs RNA $\to$ DNA with covariant derivative **U $\to$ T**:

AU $\to$ $\overset{T}{A}$;	UA $\to$ $\overset{A}{T}$;	GC $\to$ $\overset{C}{G}$;	CG $\to$ $\overset{G}{C}$;
AC $\to$ $\overset{T^*}{A}$;	CA $\to$ $\overset{A}{T^*}$;	AG $\to$ $\overset{G}{C^*}$;	GA $\to$ $\overset{C^*}{G}$;
GU $\to$ $\overset{T}{A^*}$;	UG $\to$ $\overset{A^*}{T}$;	UC $\to$ $\overset{C}{G^*}$;	CU $\to$ $\overset{G^*}{C}$;
AA $\to$ $\overset{T^*}{A^*}$;	UU $\to$ $\overset{A^*}{T^*}$;	GG $\to$ $\overset{C^*}{G^*}$;	CC $\to$ $\overset{G^*}{C^*}$;

by the following way: neighbouring pairs **AU, UA, GC, CG** form a bundle base; all other pairs will be transformed with the aid of marked point*.

Methylated DNA with isolated methylated nucleotide U* = T is identified with basic DNA.

Each marked point of DNA nucleotide has a charge of –1, unmarked point has a charge of +1. Let us arrange the charges of marked and unmarked points inside the DNA space, and then invert the upper marked points:

$$\frac{T}{A} = \begin{vmatrix} T \\ - \\ + \\ A \end{vmatrix}; \qquad \frac{G^*}{C} = \begin{vmatrix} G \\ + \\ + \\ C \end{vmatrix}; \qquad \frac{A^*}{T^*} = \begin{vmatrix} A \\ + \\ - \\ T \end{vmatrix} \text{ and so on.}$$

Let us introduce charge gauge of marked points:

A+	T–	C–	G+
T+	A–	G+	C–

and replace the space of marked points by complementary DNA pairs. Let us separately isolate the DNA of marked points and we will consider it as an intron.

For example, for RNA(0) = **UGGCCUAAUC**

```
              ACGTC     ACGTC
A*CG*T*C      +-++-     ACAGT     ACGTCACAGT
          ->        ->        ->
T GC A*G*     +++--     TGTCA     TGCAGTGTCA.
              TGCAG     TGCAG
```

Inverse transformation DNA → RNA is determined as a homeomorphism of DNA (+) strand:

ACGTC**ACAGT**
TGCAG**TGTCA** → **UGCAGUGUCA** = PHK (1).

Involutive transformation intron ↔ exon:

```
ACGTC     ACAGT     ACAGT
ACAGT     ACGTC     +-+--     A*CA*G T
      ->        ->        ->
TGTCA     TGCAG     ++--+     T GT*C*A
TGCAG     TGTCA     TGTCA
```

simply transforms nucleotides pairs of RNA into other nucleotides pairs of RNA:

UGGCCUAAUC → **UGGCUUAGAU** = PHK (2).

Performed transformations of RNA together with simple transformation **U→T** of ordinal type ω_1 contain transformations RNA nucleotide pairs of ordinal type ω_2. These transformations are non-ambiguous with fixed intron's gauge. However, transfer $\mathbf{CH_2} \rightarrow \mathbf{O}$ in RNA base is not strictly reversible because introns are production of entropy: mark **O** has the same charge $Q_p = 8^*$ as a mark $\mathbf{CH_2}$ but is only homeomorphous to it. This fact indicates that RNA has tangent Riemann space. Ordinal type ω_2 is considerably more complex than the ordering mode in classical Riemann geometry. 'Inert mass' (intron) identical to 'gravitational mass' (exon) determines the Riemann space of DNA. Cutting of introns and fixation of exons are identical in the operation of annulation of DNA–RNA bond and consequently to the annulation of RNA Riemann space. But Riemann space remains only in tangent space, which is induced by the strand of RNA basic nucleotides and not informational ones: tangent space of RNA informational nucleotides is plane four-dimensional space with a metric 2^+2^-.

Let us note that if DNA is transformed into RNA without introns cutting then RNA informational nucleotides form the Riemann tangent space, which is a copy of Riemann space of DNA informational nucleotides. Taking that into account

$$\omega_2 = \omega_1 + Q_p(\mathbf{O} \to \mathbf{CH}_2) = \omega_1 + [Q_p(2\mathbf{n})^{+2} = Q_4(\mathbf{n})],$$

RNA forms a doubled Riemann space. This situation does not have a classical analogue.

As soon as introns are cut off, the connection is induced; for example (cut introns are highlighted):

UGCAG = 1/2 PHK(1)

UG**G**CC**UA**AU**C** → UGGCUUAGAU = PHK(2) → UGCCAU**A**;

UGCCAU**A** – **informational RNA,**

and then informational RNA is the matching function in metric 2^+2^-, i.e., **negative part of connection is informational RNA**. It is not important how three connection components are placed relatively to each other, but the matching function determines the selection of that half of RNA(2), which the acceptation of informational RNA starts.

RNA(2) is a base of the RNA(0) bundle. Resulted informational RNA is projected on RNA(2) and is correlated with it. Correlation errors form poly-**A** and are attached to informational RNA. If we designate non-correlated RNA(2) nucleotides as Nu and poly-**A** as **A...A**, then

$$Q_p(\underline{\mathrm{Nu}}) = Q_p(\mathbf{A...A}).$$

Poly-**A** is attached to informational DNA and is also a negative part of the connection.

RNA(1) forms a **splicing cone**, which is **dual basis** of RNA(0) bundle.

RNA(0) introns and exons are represented by positive metric tensor: exons have a character equal to +1, intorns are equivalent to distributed terminator basis and have a character equal to –1:

UG**G**CC**UA**AU**C** → ++–++—++–.

Metric tensor of RNA(0) determines the controlling DNA, which is created as a representation of connection integration mode inside the splicing cone; informational RNA is a result of integration. In order to obtain controlling DNA, it is necessary to have a second metric tensor dual to that of RNA(0).

RNA(0) metric tensor on splicing cone acts not only on individual informational nucleotides but also on duplicated nucleotides of non-informational half of RNA(2) (Fig. 3.8).

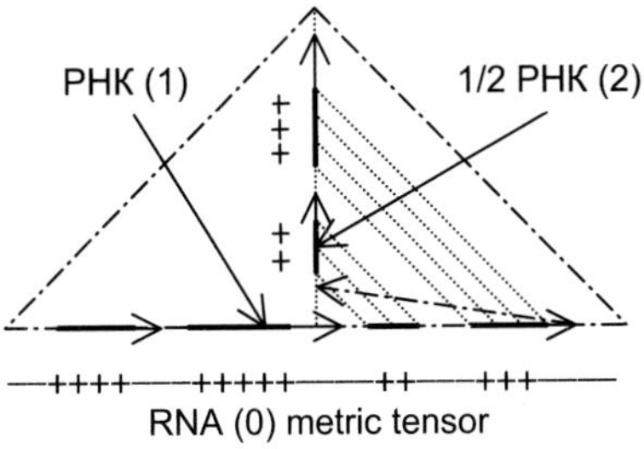

Fig. 3.8 The path of connection integration in RNA splicing cone is shown by an arrow. RNA zones corresponding to **enzyme** fragments are highlighted. **Lie group of conic screw line** with an attendant constriction of enzyme fragments and constriction of the splicing cone itself into a line, is the representation of a connection integration way. Helical and linear enzyme fragments correspond to the characters (+1) and (1) of RNA(0) metric tensor:

linear zones are non-extensible (–1) with complete enzyme stretching; helical zones are extensible (+1).

An integrated connection is identified with ferment, which is homeomorphism nucleus of controlling DNA:

```
              ++−++−−++−
PHK(1) =      UGCAGUGUCA    UG_AG__UC_
                   −−++−
1/2PHK(2) =        UAGAU  =        GA_  →  UGAGUCGA
              PHK(3)
```

Metric tensor of RNA(0) remains in an enzyme geometric structure but does not remain in the geometric structure of informational RNA.

Controlling DNA is built on the basis of RNA(3), which is a concatenation of the RNA(1) and the non-informational part of RNA(2), on which the corresponding part of RNA(0) metric tensor is transferred.

The dual metric tensor is determined by the RNA(0) metric tensor but it acts on RNA(3) lowering (+1) and raising (–1) RNA nucleotides with transformation **U → T** and filling free DNA space with complementary nucleotides.

(+)-DNA strand contains homeomorphism nucleus of RNA(0) metric tensor, which is a set of irreversible elements and is identical to the enzyme. It is clear that the enzyme is a radical of controlling DNA (genetic code transforms enzyme into protein form).

```
              ++-++--++---++-       C  TG  ATA
PHK(3) =   UGCAGUGUCAUAGAU  →  TG  AG   TC    GAT*  →

              ACCTCTGAGATACTA          Controlling
     →                             =
              TGGAGACTCTATGAT             DNA.
```

The methylation mark of a terminal thymine splits off because the terminator basis is always on the right end of controlling DNA. Informational RNA is an accepting part of primary RNA transcript and compulsorily requires the RNA splicing process to be irreversible.

Informational RNA and its enzyme form a left ring of quotients with division operator, which is RNA(0) metric tensor with proper value – informational RNA (quotient) and improper value – enzyme (remainder). Although RNA(0) metric tensor is located in geometric structure of enzyme, but only one sign (+) corresponds to the left-hand and right-hand rotation of enzyme fragments; – and this fact determines the irreversible nature of RNA(0) transcription and RNA enzyme translation.

If introns are not cut off from the RNA(0), then independently of the fact of poly-**A** attaching or non-attaching, informational RNA does not contain information on RNA(0) metric tensor but forms a new derivative of RNA(0) metric tensor, **geometric tensor** required for production of DNA, which will contain RNA(0) metric tensor:

RNA(0) geometric tensor	→ + + + – – – + + + + →	AAGCCTGAAG
RNA(0) metric tensor	→ + + – + + – – + + – →	TTCGGACTTC

in a gauge of introns. The geometric tensor is in the form of paired (–1) and (+1) individual fragments

RNA zones with uncut introns can be placed locally inside of RNA(0).

RNA enzyme production process can also be accompanied by attaching of poly-**A** and is connected with erroneous introns cutting because introns are always from RNA(3), which contains non-informational part of RNA(2).

Tables 3.1 and 3.2 contain 24 orthogonal unambiguous and 12 ambiguous gauges of introns in metric $2^+ 2^-$ of RNA internal space. DNA

Table 3.1. Unambiguous orthogonal gauges of introns

A+	T+	G–	C–	A+	T+	G–	C–	A+	T+	G–	C–
T+	A–	C+	G–	T+	A–	C–	G+	T–	A+	C+	G–
A+	T+	G–	C–	A+	T–	G+	C–	A+	T–	G+	C–
T–	A+	C–	G+	T+	A+	C–	G–	T+	A–	C–	G+
A+	T–	G+	C–	A+	T–	G+	C–	A+	T–	G–	C+
T–	A+	C+	G–	T–	A–	C+	G+	T+	A+	C–	G–
A+	T–	G–	C+	A+	T–	G–	C+	A+	T–	G–	C+
T+	A–	C+	G–	T–	A+	C–	G+	T–	A–	C+	G+
A–	T+	G+	C–	A–	T+	G+	C–	A–	T+	G+	C–
T+	A+	C–	G–	T+	A–	C+	G–	T–	A+	C–	G+
A–	T+	G+	C–	A–	T+	G–	C+	A–	T+	G–	C+
T–	A–	C+	G+	T+	A+	C–	G–	T+	A–	C–	G+
A–	T+	G–	C+	A–	T+	G–	C+	A–	T–	G+	C+
T–	A+	C+	G–	T–	A–	C+	G+	T+	A–	C+	G–
A–	T–	G+	C+	A–	T–	G+	C+	A–	T–	G+	C+
T+	A–	C–	G+	T–	A+	C+	G–	T–	A+	C–	G+

Table 3.2. Ambiguous gauges of introns

A+	T+	G–	C–	A+	T+	G–	C–	A+	T–	G+	C–
T+	A+	C–	G–	T–	A–	C+	G+	T+	A–	C+	G–
A+	T–	G+	C–	A+	T–	G–	C+	A+	T–	G–	C+
T–	A+	C–	G+	T+	A–	C–	G+	T–	A+	C+	G–
A–	T+	G+	C–	A–	T+	G+	C–	A–	T+	G–	C+
T+	A–	C–	G+	T–	A+	C+	G–	T+	A–	C+	G–
A–	T+	G–	C+	A–	T–	G+	C+	A–	T–	G+	C+
T–	A+	C–	G+	T+	A+	C–	G–	T–	A–	C+	G+

zones can be separated into domains, in which RNA information is kept in one of orthogonal introns gauges and DNA introns are kept in other orthogonal gauge. The presence of two orthogonal introns gauges in DNA determines RNA(0) metric tensor. Because RNA(3) always contains introns, DNA methylation introduces a third orthogonal gauge of introns.

Quantification of Conformal Field

4 Chapter

Quantification of field ϕ comes to the construction of tangent space at each of its point. Geometric properties of a three-dimensional tangent space are determined by metric tensor, which depends on the chemical realization of each amino acid in four-dimensional space-time. All three-index chemical configurations form a nucleus of parallel transfers to other point of the space and do not depend on the time; all 4_ϕ-index configurations depend on time and can be determined as three-index ones under condition that $\phi = 0$, i.e., in Hamiltonian gauge. Division of the set of amino acid chemical configurations on three-index and 4_ϕ-index expresses the principle of equivalency connection to space-time Lorentz-interval: each chemical configuration can be brought to standard connection form by means of introduction of marked point; for example:

$$F\{+A+P+S\} \rightarrow F\{+A-P^*+S\}$$

$$A\{+R+S-\phi-Y\} \rightarrow A\{-R^*+S-\phi-Y\}.$$

A complete set of amino acid chemical configurations as well as of connection elements G, *, ϕ forms a **plane impulse space of amino acid conformal field with diagonal metric of variable length,** negative and positive parts of which are determined by the quantity of 3- and 4_ϕ-index chemical configurations. Vector space with erased coordinates, which are introns is an analogue of this space. But precise metric can be selected as Lorentz one: one three-index and three 4_ϕ-index chemical configurations. There are no 'rigid' three-index configurations for ϕ field.

Amino acids with isolated charge A, F, S, H, P, R, W, Y and connection elements G, *, ϕ form basic 10-dimensional impulse space of CFAA bundle, each component of which is expressed through other nine components.

Amino acids with coinciding charges C, D, E, I, K, L, M, N, Q, T, V are expressed through all 10 basic components.

Operators	Functions
$\pm 1\phi$	$\pm$1**C**
$\{$ $P + * = P*$	−1**H**
$P - * = P^{-}*$	+1**H**
$\phi - * = \phi^{-}*$	**O** → **NH**
$G - \phi = -\phi*$	**O** → **NH**
$S - * \rightarrow + \mathbf{S}^{-}* - *$	+1**S** + 2**H**
$S + * = S*$	+1**S**
$W - P = W^{-P}$	**N** → **CH**
$W - P + * = W^{-P*}$	**CH** → **N**
$S - G = S^{-G}$	**CH** → **N**
$S + P + * = S^{P*}$	1**C**2**H** → **O**

Poincaré group is built on linear impulse space by means of a component pairing with generation of corresponding pairing operators. Each pairing is considered as an operator, which induces transfers between equivalent proton-charge configurations of amino acids; vector length of each configuration changes.

The heading of each element of CFAA field contains metric of tangent impulse space. Basic chemical configurations are designated as =3= or =4=. In those cases when basic configuration coincides with the genetic code configuration, amino acid symbol, * or ϕ is listed. Operator configuration is listed under basic one. Comments are introduced for explanations of complex configurations. Structural chemical formulae from set of possible structures are given in some cases.

Equivalent charge configurations are bundle over base – conditional information when preserving gauge of amino acid pairing operators. Yet, whatever is the choice of pairing operator for the simple representation of amino acid configurations, there are always equivalent charge configurations of the genetic code. We can say that it is not possible to remove conformal field of amino acids by any amino acid transformation. This statement is analogous to that that it is not possible to annihilate gravitational field by transformation of coordinates. Condition $\phi = 0$ narrows CFAA to Yang-Mills field, but does not remove equivalent charge configurations, i.e., information. Physical cause of CFAA non-destroying consists in the fact that equivalent charge configurations of amino acids have various spatial structures even with fixed gauge. Change of the gauge of amino acids pairing operators changes the energy of equivalent charge configurations and leads them to **microcanonic distribution.** Genetic code, therefore, does not depend on specific physical structure of amino acids.

The state of field ϕ with positive as well as with proper negative energy forms a discrete spectrum of values. This fact is a ground for the radical difference of CFAA from those fields, which do not take into account geometry of space-time. Quantum mechanics with operators acting in Hilbert space of wave functions, as it is known, cannot describe dynamic physical systems with discrete spectrum of positive proper values. Hilbert space of wave functions is, therefore, tangent space in each point of Riemann space of amino acids conformal field. Some analogue of field ϕ can be found in quantum systems with inverse spectrum of the proper values of energy, for example, in lasers. However, their coherent states are located in zones of negative values of atoms' energy. More exact analogue of field ϕ is the neutron-proton field of atomic nucleus. However, strong interactions of elementary particles are determined in a four-dimensional space-time and, therefore, depend on the structure of elementary particles. Conformal field of amino acids determines geometry of six-dimensional space-time and does not depend on the structure of amino acids.

Metric tensor of tangent space for each amino acid determines the quantity of field ϕ levels with positive energy (positive parts of metric tensor signature). Field ϕ does not have zero energy level. Lorentz-invariance of field ϕ requires a minimum presence of one level with negative energy and three levels with positive energy.

Three levels with positive energy of one amino acid corresponds to each level with negative energy of the triplet of basic RNA nucleotides* I25(CTC) *and* S4(TCT). *That is why the genetic code exists.

Alanine A $\{12^{+}\ 5^{-}\}$

=3=	– F + G + Y		1H	1O	

Radical **OH** is a generative radical of amino acid A.

=3=	– F – * + Y			1O	–1*
= 3 =	+ F – P – S	3C	–1H	–1O	

This configuration requires an introduction of antiprotonic states of chemical elements. It is possible to introduce two marked points: 3C **H* O***; but expansion of three-dimensional space is performed by field ϕ. This configuration is realized as field **GF**(3^3) of Lorentz-interval parallel transfer with nucleus **H**. Other configurations with CFAA marked points are examined analogically.

A	$+F+S-Y$	1C	3H				

Basic configuration of the genetic code.

=3=	$+H+P-Y$		3H	–1O	2N		
=4=	$+F+P-\phi-Y$	3C	5H	–1O			-1ϕ

Configuration 2**C** 5**H** 1**O*** corresponds to amino acid T in state $-F+P-\phi+Y$. Let us note that CFAA expansion with the aid of marked points cannot give all amino acids configurations of genetic code, for example, amino acids C, M. It is possible to introduce Yang-Mills field according to the formula:

$$\text{Am}\,(\mathbf{C,H,O,-N}) \rightarrow YM_N\,\text{Am}\,(\mathbf{C,H,O,N})$$

with the commutator –2. Then Yang-Mills fields YM_N, YM_H, YM_O and YM_C replace the marked point*.

=4=	$-G+{*}+S-\phi$	1C	2H	1O		1*	-1ϕ

For operator S^{-G}, the obstacle is a marked point*.

=4=	$+G-H-\phi+Y$	3C	3H	1O	–2N		-1ϕ

For operator $-\phi^*$, the obstacle is a marked point of chemical element **N**.

=4=	$+G-R+\phi+Y$	3C	–2H	1O	–3N		1ϕ
=4=	$-G+W-\phi-Y$	2C	4H	–1O	1N		-1ϕ
=4=	$-{*}-H-\phi+Y$	3C	2H	1O	–2N	–1*	-1ϕ
=4=	$-{*}-R+\phi+Y$	3C	–3H	1O	–3N	–1*	1ϕ
=4=	$+{*}+W-\phi-Y$	2C	5H	–1O	1N	1*	-1ϕ
=4=	$+H-P-S+\phi$		–3H	–1O	2N		1ϕ
=4=	$+H+S+\phi-Y$	–2C	1H		2N		1ϕ
=4=	$-P+R-S-\phi$		2H	–1O	3N		-1ϕ
=4=	$+R+S-\phi-Y$	–2C	6H		3N		-1ϕ

Amino acid A has only the configurations which are related to the Lorentz-interval parallel transfer; therefore, it forms an element of three-dimensional space of points, the distance between which can be determined easily. Introduction of Yang-Mills field gives an opportunity to determine the distance between different amino acids, in particular, between A and T.

But because of non-linearity of Yang-Mills field, factor-amino acids C, M are themselves generators of this field and, therefore, have to be determined with the aid of another field or by means of introduction of more complex marked points.

We can consider that amino acid A induces Yang-Mills field.

Phenylalanine F $\{8^+ \ 8^-\}$

=3=	– A + G + Y	6C	5H	1O			
=3=	– A – * + Y	6C	4H	1O			–1*
=3=	+ A + P + S	5C	11H	1O			
F	+ A – S + Y	7C	7H				

Basic configuration of the genetic code.

=3=	– G – P + W	6C	6H		1N
F	$-G + W^{-P}$	7C	7H		

Operator W^{-P} is originally introduced in order to get the precise representation of operator $\pm 1\phi = \pm 1$**C**.

=3=	+ * – P + W	6C	7H		1N	1*
	$+W^{-P*}$	5C	5H		2N	

Operator W^{-P*} is originally introduced in order to get the basic configuration of amino acid H of the genetic code.

=3=	+ H + P – S	6C	7H	–1O	2N	
=3=	– P + R + S	2C	8H	1O	3N	
=4=	+ A – P + ϕ + Y	5C	5H	1O		1ϕ
		6C	5H	1O		

When changing the gauge of selected pairing operators, this configuration can be transformed into basic configuration, for example, by introduction of new operator $+Y - P = Y^{-P}$: **O** → +1**C** + 2**H**. The state 6**C** 5**H** 1**O** is one of the excited states of amino acid F. The Lorentz-invariance of field requires the presence of two or more 4_ϕ-states of amino acid F, for example,

=4=	– G – S + W – ϕ	8C	8H	–1O	1N	–1ϕ
		7C	8H	+1O*	1N	

=4=	+ G + * + R − φ	4C	11H		3N	1*	−1φ
		3C	10H*		3N		

=4=	+ G + * + H + φ	4C	6H		2N	1*	1φ
=4=	− G − * + H + φ	4C	4H		2N	−1*	1φ
=4=	− G − * + R − φ	4C	9H		3N	−1*	−1φ
=4=	− G + * − φ + Y	7C	6H	1O		1*	−1φ
=4=	+ * − S + W − φ	8C	9H	−1O	1N	1*	−1φ

Glycine G {16$^+$ 5$^-$}

=3=	+ A + F − Y	1C	3H	−1O	
=3=	− F − P + W	−1C			1N
G	− F + W^{-P}		1H		

Basic configuration of the genetic code.

=3=	+ * − R + Y	3C	−3H	1O	−3N	1*	
=3=	− R − S + W	4C	−1H	−1O	−2N		
=3=	+ S − W + Y	−1C	−2H	2O	−1N		
=4=	− A + * + S − φ			1O		1*	−1φ
=4=	+ A + H + φ − Y	−2C	1H	−1O	2N		1φ
=4=	+ A + R − φ − Y	−2C	6H	−1O	3N		−1φ
=4=	− A + W − φ − Y	1C	2H	−1O	1N		−1φ
=4=	+ F − * − H − φ	3C	2H		−2N	−1*	−1φ
=4=	− F − * + H + φ	−3C	−2H		2N	−1*	1φ
=4=	+ F − * − R + φ	3C	−3H		−3N	−1*	1φ
=4=	− F − * + R − φ	−3C	3H		3N	−1*	−1φ
=4=	− F + * − φ + Y			1O		1*	−1φ
=4=	− F − S + W − φ	1C	2H	−1O	1N		−1φ
=4=	− * + P − S − φ	2C	2H	−1O		−1*	−1φ
S*	+ P^{-*} − S − φ	1C	3H	+1O*			
=4=	− * − P + S + φ	−2C	−2H	1O		−1*	1φ
=4=	− H − P + W − φ	2C	2H		−1N		−1φ
G	− H + W^{-P} − φ		1H				
=4=	− H + R − S + φ	−1C	2H	−1O	1N		1φ
=4=	− P − R + W + φ	2C	−3H		−2N		1φ
=4=	+ P − W − φ + Y	1C		1O	−1N		−1φ

Terminator* {16⁺ 5⁻}

The terminator is equivalent to antiproton. The terminator has inverse metric: 16^+ basic states of chemical element **S** and 5 excited states. Chemical elements **C, H, O, N, P** are excited states of the terminator:

$$Q_p\{\mathbf{C,H,O,N,P},*\} = \phi^{+2}$$

Field ϕ^{+2} is a copy of field ϕ.

=3=	– A – F + Y	–1C	–3**H**	1O
S*		+1C*	+3**H**	1O

This terminator mark induces pairing operators of amino acids S* and $\mathbf{S}^{-*}$ when expanding ϕ.

Let us introduce the field M_* of a complex marked point of terminator bound with the three operators:

$$M_* = \begin{cases} S - * \rightarrow + \mathbf{S}^- * - * & +1\mathbf{S} + 2\mathbf{H} \\ S + * = S* & +1\mathbf{S} \\ \mathbf{S}* = \mathbf{P} & +1\mathbf{P}, \end{cases}$$

where **P** is a chemical element $^{31}\mathbf{P}_{15}$:

$$Q_p\{+1\mathbf{S} + 2\mathbf{H}, +1\mathbf{S}, +1\mathbf{P}\} = F(49).$$

Charge configuration with chemical element **P**, for example, **P**-treonine (2**C** 6**H** 4**O** 1**P**)** can be interpreted as a configuration 2**C** –5**H*** 4**O** –1**S*** of amino acid P*(23). Indeed, if we select operator S + * as S + * → + S* –*: +1**P** then, **P** = **S*** + *: +1**S** – 2**H** and field M_{-*} is symmetric:

$$M_{-*} = \begin{cases} S - * \rightarrow + \mathbf{S}^- * - * & +1\mathbf{S} + 2\mathbf{H} \\ S + * \rightarrow + \mathbf{S}* - * & +1\mathbf{P} \\ \mathbf{P}* = \mathbf{S}* + * & +1\mathbf{S} - 2\mathbf{H}. \end{cases}$$

$$Q_p\{+1\mathbf{S} + 2\mathbf{H}, +1\mathbf{P}, +1\mathbf{S} - 2\mathbf{H}\} = \mathbf{GF}(47).$$

Commutator $[M_*, M_{-*}] = -2$. Operator M_* introduces asymmetry in charge configurations of the genetic code with the advantage of **S** over **P**.

---

*	+ F + P – W	1C	–1**N**

The terminator mark belongs to the amino acid P when breaking into pirrolydine ring.

=3=	+ G + R - Y	-3C	4H	–1O	3N	
=3=	+ R + S - W	-4C	1H	1O	2N	
=3=	- S + W - Y	1C	2H	–2O	1N	
=4=	+ A + G - S + ϕ		1H	–1O		1ϕ
*		1C	1H	–1O		

This terminator mark belongs to complementary pair **AT** when breaking hydrogenous bonds **O…H…C**.

=4=	– A – H – ϕ + Y	2C	–1H	1O	-2N	–1ϕ
=4=	– A – R + ϕ + Y	2C	– 6H	1O	– 3N	1ϕ
=4=	+ A – W + ϕ + Y	–1C	– 2H	1O	– 1N	1ϕ
=4=	+ F – G – H – ϕ	3C	1H		– 2N	–1ϕ
=4=	– F – G + H + ϕ	–3C	–3H		2N	1ϕ
=4=	+ F – G – R + ϕ	3C	– 4H		– 3N	1ϕ
=4=	– F – G + R – ϕ	–3C	2H		3N	–1ϕ
=4=	+ F + G + ϕ – Y		1H	–1O		1ϕ
*		1C	1H	–1O		
=4=	+ F + S – W + ϕ	–1C	– 2H	1O	– 1N	1ϕ
=4=	– G + P – S – ϕ	2C	1H	–1O		–1ϕ
*		1C	1H	–1O		
=4=	– G – P + S + ϕ	–2C	– 3H	1O		1ϕ
=4=	+ H + P – W + ϕ	–2C	– 2H		1N	1ϕ
=4=	+ H – R + S – ϕ	1C	– 2H	1O	–1N	–1ϕ
=4=	+ P + R – W – ϕ	–2C	3H		2N	–1ϕ
=4=	– P + W + ϕ – Y	–1C		–1O	1N	1ϕ
*	+ W^{-P} + ϕ – Y	1C	1H	–1O		

Terminator mark, which is used during differentiation of amino acids I, L.

Histidine H {11^{+} 2^{-}}

=3=	+ A - P + Y	5C	5H	1O
=3=	+ F - P + S	5C	5H	1O

These two configurations are radical of DNA furanose ring and are determined as an ideal of U_{-2} duplication of DNA basic nucleotides (Fig. 4.1).

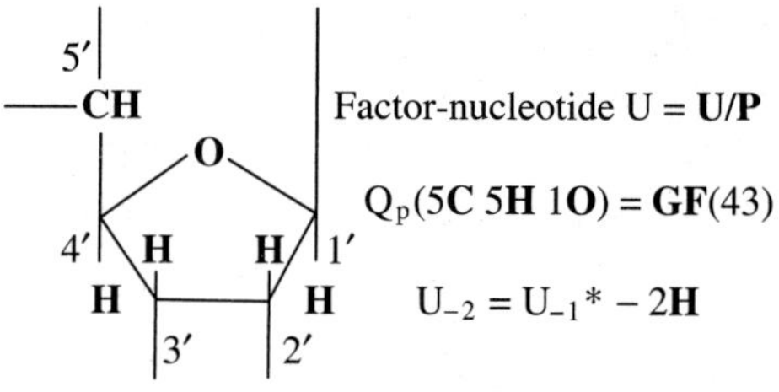

Fig. 4.1. Furanose ring radical U = **U/P** has five free valence bonds; it is dual to phosphate basis of nucleotide C.

Let us introduce operators of furanose ring:

$$FU^{+} = C = U_{-1}^{*} + 2\mathbf{H};$$
$$FU_{0} = U_{-1}^{*};$$
$$FU_{-} = U_{-2} = U_{-1}^{*} - 2\mathbf{H} = \mathbf{U/P}.$$

Operator FU^{+} contains the proper chemical element **P**; operators FU^{+}, FU_{0} form $\mathbf{Z}_2$ graduation of DNA basis.

=4=	– A + G – ϕ + Y	6C	5H	1O			–1ϕ
	– A – ϕ* + Y	5C	6H		1N		
=4=	– A – * – ϕ + Y	6C	4H	1O		–1*	–1ϕ
=4=	+ A + P + S – ϕ	5C	11H	1O			–1ϕ
		4C	11H	1O			
=4=	+ A – S – ϕ + Y	7C	7H				–1ϕ
		6C	7H				
=4=	+ F + G + * – ϕ	7C	8H			1*	–1ϕ
=4=	+ F – G – * – ϕ	7C	6H			–1*	–1ϕ
=4=	– G – P + W – ϕ	6C	6H		1N		–1ϕ
	– G + W^{-P} – ϕ	6C	7H				
=4=	– G + R – S + ϕ	3C	6H	–1O	3N		1ϕ
=4=	+ * – P + W – ϕ	6C	7H		1N	1*	–1ϕ
H	+ W^{-P*} – ϕ	4C	5H		2N		

Basic configuration of the genetic code.

=4=	+ * + R – S + ϕ	3C	7H	–1O	3N	1*	1ϕ
=4=	– P + R + S – ϕ	2C	8H	1O	3N		–1ϕ
		1C	8H	1O	3N		

Proline P* {12⁺ 6⁻}

=3=	– A + F – S	5C	1H	–1O			
=3=	+ A – H + Y	4C	5H	1O	–2N		
=3=	– F – G + W	2C	4H		1N		
=3=	– F + * + W	2C	5H		1N	1*	
=3=	+ F – H + S	4C	5H	1O	–2N		
=3=	– F + R + S	–2C	6H	1O	3N		
=4=	+ A – F + ϕ + Y	1C	3H	1O			1ϕ
		2C	3H	1O			

Aspargine semi-aldehyde **$OHCHH_2$** –

=4=	– A + H – S + ϕ	2C	–1H	–1O	2N		1ϕ
=4=	– A + R – S – ϕ	2C	4H	–1O	3N		–1ϕ
=4=	+ G + * + S + ϕ	1C	4H	1O		1*	1ϕ
	+ G + S* + ϕ	1C	1H				1S
=4=	– G – * + S + ϕ	1C	2H	1O		–1*	1ϕ
	+ S^{-G} + ϕ^{-*}	1C	3H		2N		
	– G + $\mathbf{S}^{-*}$ + ϕ^{-*}	1C	1H				1S

Operator + S – * in this configuration acts after operator ϕ – *. Radical –**CH** = **S** is primary proline radical and contains basis of terminator **S**.

=4=	– H + R + S – ϕ	1C	8H	1O	1N	–1ϕ
			8H	1O	1N	

Basic configuration of the genetic code.

In order to obtain proper marked point of amino acid P*, the transfer: –1**O** → – 1**N** – 1**H** is necessary. Then terminator basis increases up to M(41)* = + 1P + 1**S** – 1**H*** (each element of this identity has a proper marked point).

=4=	- H + R + S - ϕ	1C	8H	1O	1N	-1ϕ
			8H	1O	1N	

This configuration is one of those where the chemical valence of atoms is lower than their possible associative connection as determined by the proton charge of each atom.

Arginine R {11⁺ 4⁻}

=3=	+ F + P − S	9C	9H	−1O			
=3=	− G + * + Y	7C	6H	1O		1*	
=3=	− G − S + W	8C	8H	−1O	1N		
=3=	+ * − S + W	8C	9H	−1O	1N	1*	
=4=	− A + G + ϕ + Y	6C	5H	1O			1ϕ
		7C	5H	1O			
=4=	− A − * + ϕ + Y	6C	4H	1O		−1*	1ϕ
	− A + ϕ^{-*} + Y	7C	6H		1N		
=4=	+ A + P + S + ϕ	5C	11H	1O			1ϕ
		6C	11H	1O			
=4=	+ A − S + ϕ + Y	7C	7H				1ϕ
		8C	7H				
=4=	+ F + G + * + ϕ	7C	8H			1*	1ϕ
=4=	+ F − G − * + ϕ	7C	6H			−1*	1ϕ
=4=	+ G + H + S − ϕ	5C	9H	1O	2N		−1ϕ
		4C	9H	1O	2N		

This configuration corresponds to citrulline:

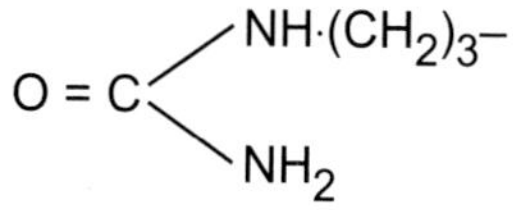

R	+ H + S − ϕ*	4C	10H		3N		

Basic configuration of the genetic code.

=4=	− G − P + W + ϕ	6C	6H		1N		1ϕ
	− G + W^{-P} + ϕ	8C	7H				
=4=	− * + H + S − ϕ	5C	8H	1O	2N	−1*	−1ϕ
	− * + H + $\mathbf{S}^{-*}$ − ϕ	3C	7H		2N		1S
=4=	+ * − P + W + ϕ	6C	7H		1N	1*	1ϕ
	+ W^{-P*} + ϕ	6C	5H		2N		
=4=	+ H + P − S + ϕ	6C	7H	−1O	2N		1ϕ

Serine S {12^+ 8^-}

Amino acid S is the bond of field **GF**(83) of the genetic code: S = H(43) + Y(57) – **GF**(83). Therefore, all 20 amino acids can be represented via the states of amino acid S. Amino acids with isolated charge form basis 8^- of impulse space of amino acid S (mass part); all amino acids with coinciding charges are states excited by field φ. Field **GF**(17) transfers itself.

=3=	– A + F – P	3C	–1H	
S	+ A – F + Y	1C	3H	1O

Basic configuration of the genetic code.

=3=	– F + H + P		3H		2N		
=3=	+ F + P – R	6C	2H		–3N		
=3=	– G – R + W	5C	1H		–2N		
=3=	+ G + W – Y	2C	6H	–1O	1N		
=3=	+ * – R + W	5C	2H		–2N	1*	
=3=	– * + W – Y	2C	5H	–1O	1N	–1*	
=4=	+ A + G – * + φ	1C	4H			–1*	1φ
=4=	– A + H – P + φ		– 3H		2N		1φ
=4=	+ A – H – φ + Y	4C	5H	1O	–2N		–1φ
=4=	– A – P + R – φ		2H		3N		–1φ
=4=	+ A – R + φ + Y	4C		1O	–3N		1φ
=4=	– F – G + W – φ	2C	4H		1N		–1φ
		1C	4H		1N		
=4=	– F + * + W – φ	2C	5H		1N	1*	–1φ
=4=	+ G + * + P – φ	3C	6H			1*	–1φ
	+ G + P* – φ	2C	5H				
=4=	– G – * + P – φ	3C	4H			–1*	–1φ
	– G + P^{-*} – φ	2C	5H				

2-aminobutyric acid $\mathbf{CH_3CH_2}$ –

=4=	– G – H + R + φ		4H		1N		1φ
		1C	4H		1N		
=4=	+ * – H + R + φ		5H		1N	1*	1φ
=4=	+ H + P – R + φ	3C			–1N		1φ

Let us prove that amino acid S has the proper marked point. To do it, let us build a triangle of differences $Q_p(Am) - Q_p(S)$ of amino acids with isolated charge:

W*

−40 +32

+28 −12 +20

−18 +10 −2 +18

+6 −12 −2 −4 +14

+14	+20	$+6_{(+2)}$ =	+8	+6	+2	+16
−8	+6	+26	+32	+38	+40	+56
A	P	H	F	R	Y	W
9	23	43	49	55	57	73

When calculating differences, it is necessary to account for coincidence of charges: each coinciding charge gets additional increment +2, +3,... in the direction opposite to the ordering direction.

Terminator mark with basis **S** belonging to amino acid S was transferred to amino acid W with terminator basis **H**, which was to be proved.

Tryptophan W {$10^+ 6^-$}

=3=	+ F + G + P	10C	13H				
=3=	+ F − * + P	10C	12H			−1*	
	+ F + P^{-*}	10C	13H				
=3=	+ G + R + S	5C	14H	1O	3N		
=3=	− G + S + Y	8C	9H	2O			
	+ S^{-G} + Y	7C	8H	2O	1N		
=3=	− * + R + S	5C	13H	1O	3N	−1*	
	− * + R + **S**$^{-*}$	4C	12H	3N			1S
=3=	+ * + S + Y	8C	10H	2O		1*	
	S*+ Y	7C	7H	1O			1S
=4=	+ A + G + ϕ + Y	8C	11H	1O			1ϕ
		9C	11H	1O			
=4=	+ A − * + ϕ + Y	8C	10H	1O		−1*	1ϕ
W	+ A + $ϕ^{-*}$ + Y	9C	12H		1N		

Basic configuration of the genetic code.

=4=	+ F + G + S + ϕ	8C	11H	1O			1ϕ
		9C	11H	1O			
=4=	+ F − * + S + ϕ	8C	10H	1O		−1*	1ϕ
	+ F − * + **S**¯* + ϕ	8C	9H				1**S**
=4=	+ G + H + P + ϕ	7C	11H		2N		1ϕ
		8C	11H		2N		
=4=	+ G + P + R − ϕ	7C	16H		3N		−1ϕ
		6C	16H		3N		
=4=	− G + P − ϕ + Y	10C	11H	1O			−1ϕ
		9C	11H	1O			
=4=	− * + H + P + ϕ	7C	10H		2N	−1*	1ϕ
	+ H + P¯* + ϕ	8C	11H		2N		
=4=	− * + P + R − ϕ	7C	15H		3N	−1*	−1ϕ
	+ P¯*+R − ϕ	6C	16H		3N		
=4=	+ * + P − ϕ + Y	10C	12H	1O		1*	−1ϕ
	+ P* − ϕ + Y	9C	11H	1O			

Let us build the triangle of differences for this amino acid.

						− P¯* +*						
					+2		−23					
				+9		+11		−12				
			−10		−1		+10		−2			
		−1		−11		−12		−2		−4		
	$+8_{(+2)}$		$+6_{(+3)}$		+20		$+6_{(+2)}$		+6		+2	
−64		−56		−50		−30		−24		−18		−16
A		S		P		H		F		R		Y
9		17		23		43		49		55		57

Marked point of amino acid W with basis **H** was transferred to amino acid M(41) with basis **GF**(4^2) + **GF**(5^2) = **GF**(41). Therefore, amino acid P contains the marked point of terminator with basis 40.

Field Basis ϕ $\{30^{+}\ 0^{-}\}$

Basic configurations of field ϕ are expressed unequivocally only through amino acids with isolated charge and connection elements G and *. Therefore the metric tensor of base of field ϕ does not contain a negative part. Any action of pairing operators of amino acids generates field ϕ. For example, operator W – P = W^{-P} with function **N** → **CH** transforms a basic configuration of amino acids –G –H –P +W to field ϕ with base **1C**. If in addition to introduce operator P – W = $-W^{-P}$ with function 1**N** –1**C** → **H** then negative part will appear in metric tensor of base of field ϕ.

=4=	− A + F + P − Y	2C	2H	−1O			
=4=	− A − G + * + S		−1H	1O		1*	
ϕ	− A − G + S*	−1C	−4H				1S
=4=	− A + G − H + Y	2C		1O	−2N		
=4=	+ A − G + R − Y	−2C	5H	−1O	3N		
=4=	− A − G + W − Y	1C	1H	−1O	1N		
=4=	− A − * − H + Y	2C	−1H	1O	−2N	−1*	
=4=	+ A + * + R − Y	−2C	6H	−1O	3N	1*	
=4=	− A + * + W − Y	1C	2H	−1O	1N	1*	
=4=	+ A − H + P + S	1C	6H	1O	−2N		
=4=	+ A − H − S + Y	3C	2H		−2N		
=4=	− A − P + R − S	−1C	−1H	−1O	3N		
=4=	− A + R + S − Y	−3C	3H		3N		
=4=	+ F + G + * − H	3C	3H		−2N	1*	
=4=	+ F − G − * − H	3C	1H		−2N	−1*	
=4=	− F + G + * + R	−3C	4H		3N	1*	
=4=	− F − G − * + R	−3C	2H		3N	−1*	
=4=	− F − G + * + Y		−1H	1O		1*	
=4=	− F − G − S + W	1C	1H	−1O	1N		
=4=	− F + * − S + W	1C	2H	−1O	1N	1*	
=4=	+ G + * + P − S	2C	3H	−1O		1*	
ϕ	+ G + P* − S	2C	2H	−1O			
=4=	− G − * + P − S	2C	1H	−1O		−1*	
ϕ	− G + P^{-*} − S	2C	2H	−1O			
=4=	− G − H − P + W	2C	1H		−1N		
ϕ	− G − H + W^{-P}	1C					
=4=	+ G + H − R + S	1C	−1H	1O	−1N		
=4=	+ G + P + R − W	−2C	4H		2N		
=4=	− G + P − W + Y	1C	−1H	1O	−1N		

=4=	+ * − H − P + W	2C	2H		−1N	1*	
φ	− H + W^{-P*}	1C					
=4=	− * + H − R + S	1C	−2H	1O	−1N	−1*	
φ	− * + H − R + **S**$^{-*}$		−3H		−1N		1S
=4=	− * + P + R − W	−2C	3H		2N	−1*	
φ	P^{-*} + R − W	−2C	4H		2N		
=4=	+ * + P − W + Y	1C		1O	−1N	1*	
φ	+ P* − W + Y	1C	−1H	1O	−1N		
=4=	− H − P + R + S	−2C	3H	1O	1N		

Tyrosine Y {12^+ 7^-}

=3=	+ A + F − G	8C	9H			
=3=	+ A + F + *	8C	10H			1*
Y	− A + F + S	7C	7H	1O		

Basic configuration of the genetic code.

=3=	− A + H + P	6C	7H		2N		
=3=	+ G − * + R	4C	11H		3N	−1*	
=3=	+ G − S + W	8C	10H	−1O	1N		
=3=	− * − S + W	8C	9H	−1O	1N	−1*	
=4=	− A + F + P − φ	9C	9H				−1φ
		8C	9H				
=4=	+ A − G + H + φ	5C	7H		2N		1φ
		6C	7H		2N		
=4=	+ A − G + R − φ	5C	12H		3N		−1φ
		4C	12H		3N		
=4=	− A − G + W − φ	8C	8H		1N		−1φ
		7C	8H		1N		
=4=	+ A + * + H + φ	5C	8H		2N	1*	1φ
=4=	+ A + * + R − φ	5C	13H		3N	1*	−1φ
=4=	− A + * + W − φ	8C	9H		1N	1*	−1φ
=4=	− A + H + S + φ	4C	5H	1O	2N		1φ
		5C	5H	1O			
=4=	− A + R + S − φ	4C	10H	1O	3N		−1φ
		3C	10H	1O	3N		
=4=	+ F + G − * + φ	7C	8H			−1*	1φ
=4=	+ G − P + W + φ	6C	8H		1N		1φ
	+ G + W^{-P} + φ	8C	9H				
=4=	− * − P + W + φ	6C	7H		1N	−1*	1φ

Let us build the triangle of differences of amino acid Y:

A*

+2 −7

+9 +11 +4

−10 −1 +10 +14

−1 −11 −12 −2 +12

$+8_{(+2)}$ $+6_{(+3)}$ +20 $+6_{(+2)}$ +6 +18

−48	−40	−34	−14	−8	−2	+16
A	S	P	H	F	R	W
9	17	23	43	49	55	73

Difference operator $Y - A^* = 64^{-**} = Y_{(+2)} + 1N$ induces transition **O** → 8**H** while sum operator of DNA transcription – thymine **T** = I + L induces inverse transition 8**H** → **O**. Since W = **T** + 1N* then we have a chain of transitions $Y_{(+2)} \rightarrow$ **T** → W of field ϕ during movement of synchronous basis of finite projected plane P(2) of DNA reading. It follows from here that **operator Y** transfers basic nucleotide **T** into amino acid W and therefore, is **linear predictor**. Nucleotide **T** is a marked point of amino acid Y.

Thus, operator Y is exact operator.

Cysteine, Threonine, Valine C, T, V {20⁺ 14⁻}

All 20^+ excited states of field **GF**(5^2) determine amino acids of the genetic code **GF**(Am) in field **GF**(83); 14^- basic states correspond to charges Q_p(Am).

=3=	+ A − G + S	2C	5H	1O			
	$+ A + S^{-G}$	1C	4H	1O	1N		
=3=	+ A + * + S	2C	6H	1O		1*	
C	+ A + S*	1C	3H				1S

Basic configuration of the genetic code.

=3=	− A − P + Y	3C	−1H	1O	
=3=	+ A + W − Y	3C	8H	−1O	1N
=3=	+ F − G − P	4C	1H		
=3=	− F + G + W	2C	6H		1N

2,4-diaminbutyric acid $H_2NCH_2CH_2-$

=3=	+ F + * − P	4C	2H	1*

=3=	− F − * + W	2C	5H		1N	−1*	
=3=	− F + S + Y	1C	3H	2O			
=3=	+ G − * + P	3C	6H			−1*	
V	+ G + P^{-*}	3C	7H				

Basic configuration of the genetic code.

=3=	− G + H − S	3C	1H	−1O	2N		
=3=	+ * + H − S	3C	2H	−1O	2N	1*	
=3=	+ H + R − W	−1C	3H		4N		
=3=	+ P − R + Y	6C	2H	1O	−3N		
=4=	+ A − G + P − ϕ	4C	7H				−1ϕ
V		3C	7H				
=4=	+ A + * + P − ϕ	4C	8H			1*	−1ϕ
V	+ A + P* − ϕ	3C	7H				
=4=	− A + P + S − ϕ	3C	5H	1O			−1ϕ
T		2C	5H	1O			

Basic configuration of the genetic code.

=4=	− A − S − ϕ + Y	5C	1H				−1ϕ
		4C	1H				
=4=	+ F − G − S − ϕ	6C	3H	−1O			−1ϕ
=4=	+ F + * − S − ϕ	6C	4H	−1O		1*	−1ϕ
=4=	+ F + H − W + ϕ	2C			1N		1ϕ
		3C			1N		
=4=	− F + P − ϕ + Y	3C	5H	1O			−1ϕ
T		2C	5H	1O			

Homoserine **$HOCH_2CH_2$** −

=4=	+ F + R − W − ϕ	2C	5H		2N		−1ϕ
		1C	5H		2N		
=4=	+ G − * + S + ϕ	1C	4H	1O		−1*	1ϕ
	+ G + S^{-*} + ϕ^{-*}		2H		1N		1S
=4=	− G + H − P + ϕ	1C	−1H		2N		1ϕ
=4=	+ G − H + W − ϕ	5C	8H		−1N		−1ϕ

=4=	−G−P+R−φ	1C	4H		3N		−1φ
			4H		3N		
=4=	+G−R+W+φ	5C	3H		−2N		1φ
=4=	+*+H−P+φ	1C			2N	1*	1φ
=4=	−*−H+W−φ	5C	7H		−1N	−1*	−1φ
=4=	+*−P+R−φ	1C	5H		3N	1*	−1φ
=4=	−*−R+W+φ	5C	2H		−2N	−1*	1φ
=4=	−H+S−φ+Y	4C	5H	2O	−2N		−1φ
=4=	−R+S+φ+Y	4C		2O	−3N		1φ

Aspartic Acid, Asparagine D, N $\{19^{+}\ 13^{-}\}$

All 19^{+} excited states of field **GF**(31) determine amino acids of genetic code **GF**(Am) with the exception of glycine G(1) in field **GF**(83); 13^{-} basic states of field **GF**(31) correspond to charges Q_p(Am) with the exception of glycine G(1).

=3=	+A−G+P	4C	7H		
=3=	+A+*+P	4C	8H		1*
	+A+P*	4C	7H		
=3=	−A+P+S	3C	5H	1O	

This configuration is a radical of amino acid hydroxyproline:

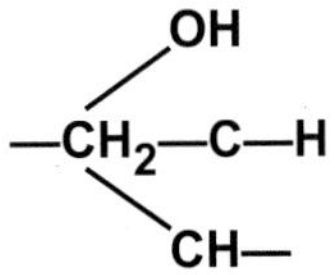

=3=	−A−S+Y	5C	1H			
=3=	+F−G−S	6C	3H	−1O		
=3=	+F+*−S	6C	4H	−1O		1*
=3=	−F+P+Y	3C	5H	1O		
=3=	+F+R−W	2C	5H		2N	
=3=	+G−H+W	5C	8H		−1N	
=3=	−G−P+R	1C	4H		3N	
=3=	−*−H+W	5C	7H		−1N	−1*
=3=	+*−P+R	1C	5H		3N	1*
=3=	−H+S+Y	4C	5H	2O	−2N	

=4=	+ A − G + S + ϕ	2C	5H	1O			1ϕ
N	+ A + S^{-G} + ϕ	2C	4H	1O	1N		

Basic configuration of the genetic code.

=4=	+ A + * + S + ϕ	2C	6H	1O		1*	1ϕ
	+ A + S* + ϕ	2C	3H				1S
=4=	− A − P + ϕ + Y	3C	−1H	1O			1ϕ
=4=	+ A + W + ϕ − Y	3C	8H	−1O	1N		1ϕ
=4=	+ F − G − P + ϕ	4C	1H				1ϕ
		5C	1H				
=4=	− F + G + W + ϕ	2C	6H		1N		1ϕ
		3C	6H		1N		
=4=	+ F + * − P + ϕ	4C	2H			1*	1ϕ
=4=	− F − * + W + ϕ	2C	5H		1N	−1*	1ϕ
=4=	+ F + H − R − ϕ	7C	2H		−1N		−1ϕ
=4=	− F + S + ϕ + Y	1C	3H	2O			1ϕ
D		2C	3H	2O			

Basic configuration of the genetic code.

=4=	+ G − * + P + ϕ	3C	6H			−1*	1ϕ
	+ G + P^{-*} + ϕ	4C	7H				
=4=	− G + H − S + ϕ	3C	1H	−1O	2N		1ϕ
=4=	− G + R − S − ϕ	3C	6H	−1O	3N		−1ϕ
=4=	+ * + H − S + ϕ	3C	2H	−1O	2N	1*	1ϕ
=4=	+ * + R − S − ϕ	3C	7H	−1O	3N	1*	−1ϕ
=4=	+ H − P + S − ϕ	2C	3H	1O	2N		−1ϕ
		1C	3H	1O	2N		
=4=	− H + P − ϕ + Y	6C	7H	1O	−2N		−1ϕ
=4=	+ H + R − W + ϕ	−1C	3H		4N		1ϕ
		3H		4N			
=4=	+ P − R + ϕ + Y	6C	2H	1O	−3N		1ϕ

Glutamic Acid, Glutamine E, Q {18^{+} 11^{-}}

Basic states 11^{-} belong to amino acids with coinciding charges

(C, T, V), (D, N), (I, L), (E, Q), (K, M),

which are the generators of field **GF**(83):

$$Q_p\{+C+T+V-D-N+I+L-E-Q+K+M\} = \mathbf{GF}(83).$$

Excited states 18^+ determine amino acids of the genetic code with the exception of E, Q.

Charge Q(39) is located in module Q(35):

$$Q(35) - 2\mathbf{GF}(3^2)^+ = Q(39) + 2\mathbf{GF}(11)^- = S(17).$$

Therefore, amino acid S has the proper marked point **S**.

Charge Q(39) bundle (trivial, as protein basis) over Q(35):

$$Q(35) = \mathbf{5} \times 7^- \supset Q(39) = \mathbf{3} \times 13^+$$

induces ordering according to type **5' 3'** as ordering of CFAA states:

$$\begin{array}{cccccc} Av & A1 & A2 & A3 & A2' & A3' \\ \partial\mathbf{Q}5 = & \mathbf{5'} = & \mathbf{Q}4 = & \partial\mathbf{Q}3 = & \mathbf{3'} = & \mathbf{Q}1 \end{array}$$

$$\mathbf{Q}2 \quad = \quad \mathbf{Q}3$$

We can consider that **right charge** Q(39) in a four-dimensional space with metric $3^+ 1^-$ turns into **left charge** Q(35) in a six-dimensional space with metric $4^+ 2^-$:

$$\partial\mathbf{Q}2 = 3^+ 1^- + 4^+ 2^- = \mathbf{Q}4.$$

This relation shows that difference between field ϕ and its copy comes to the difference between charges $Q_p(Am)$ and $Q_4 = h(Am)Q_p(Am)$.

Amino acid Q is the only protein basis. Since $\mathbf{GF}(29) = Q(39) / \mathbf{CH_4}$, then S–S bridges of C–C pairing close storing contour of field ϕ, which is required for duplication of amino acid M(41) (Fig. 4.2).

=3=	$-A+F-G$	6C	3H			
=3=	$-A+F+*$	6C	4H			1*
=3=	$+A-H+W$	6C	10H		−1N	
=3=	$-G+P+S$	4C	7H	1O		
Q	$+P+S^{-G}$	3C	6H	1O	1N	

Basic configuration of the genetic code.

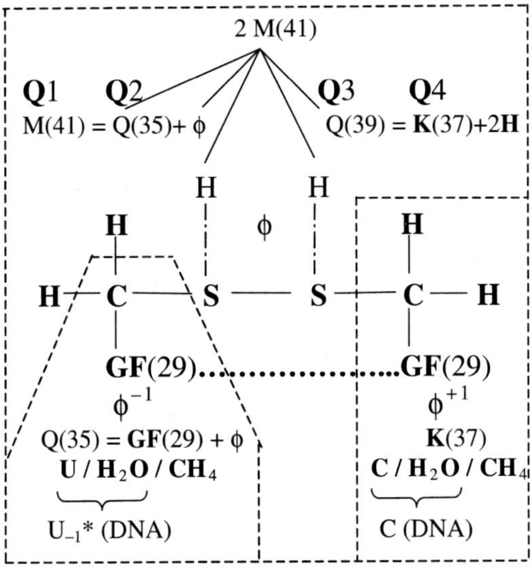

Fig. 4.2 Storing contour of field ϕ for duplication of mark M(41)

	+ S^{-P}	2C	4H	2O	1N

Introduction of recurring operator $P + S^{-G} = S^{-P}$: 1C2H → O projects amino acid Q on basic configuration –$CH(NH_2)COOH$ of amino acid.

=3=	+ G + R – S	3C	8H	–1O	3N		
=3=	– G – S + Y	6C	3H				
=3=	+ * + P + S	4C	8H	1O		1*	
E	+ S^{P*}	3C	5H	2O			

Basic configuration of the genetic code.

=3=	– * + R – S	3C	7H	–1O	3N	–1*	
=3=	+ * – S + Y	6C	4H			1*	
=3=	+ P + W – Y	5C	10H	–1O	1N		
=3=	+ R – W + Y	2C	5H	1O	2N		
=4=	+ A – F + W + ϕ	3C	8H		1N		1ϕ
		4C	8H		1N		
=4=	– A – G + H + ϕ	3C	1H		2N		1ϕ
		4C	1H		2N		
=4=	+ A + G + P + ϕ	4C	9H				1ϕ
		5C	9H				

=4=	– A – G + R – φ	3C	6H		3N		–1φ
		2C	6H		3N		
=4=	– A + * + H + φ	3C	2H		2N	1*	1φ
=4=	+ A – * + P + φ	4C	8H			–1*	1φ
	+ A + P^{-*} + φ	5C	9H				
=4=	– A + * + R – φ	3C	7H		3N	1*	–1φ
=4=	+ F + G – S + φ	6C	5H	–1O			1φ
=4=	+ F – * – S + φ	6C	4H	–1O		–1*	1φ
=4=	+ F – W + φ + Y	5C	2H	1O	–1N		1φ
=4=	+ G – * + H – φ	4C	6H		2N	–1*	–1φ
=4=	+ G – P + R + φ	1C	6H		3N		1φ
		2C	6H		3N		
=4=	– G – P + φ + Y	4C	1H	1O			1φ
		5C	1H	1O			
=4=	– * – P + R + φ	1C	5H		3N	–1*	1φ
=4=	+ * – P + φ + Y	4C	2H	1O		1*	1φ
=4=	+ H – R – φ + Y	7C	2H	1O	–1N		–1φ
=4=	– P – S + W + φ	5C	4H	–1O	1N		1φ
=4=	+ S + W + φ – Y	3C	8H		1N		1φ
		4C	8H		1N		

Leucine, Isoleucine L, I {19⁺ 14⁻}

=3=	+ A – F + W	3C	8H	1N

Ornitine $H_2NCH_2CH_2CH_2$ –

=3=	– A – G + H	3C	1H	2N
L	+ A + G + P	4C	9H	

Basic configuration of the genetic code of ordinal type ω_3.

=3=	– A + * + H	3C	2H	2N	1*
=3=	+ A – * + P	4C	8H		–1*
I	+ A + P^{-*}	4C	9H		

Basic configuration of the genetic code of ordinal type ω_2.

Let us **replace the marked point by equivalent configuration** * = + W^{-P} + φ – Y for ordinal type ω_2.

Then,

$$L/(A+P) = G; I/(A+P) = Y - W^{-P} - \phi; A+P = 32^{**}.$$

It follows from this statement that amino acids L and I are dual in a finite projected plane P(2) and, consequently, the charges **Q2** and **Q3** are the incident charges (Fig. 4.3).

$$
\begin{array}{ll}
& \quad\quad\quad\quad\quad \textbf{Q4} \\
\text{P + A:} & - \textbf{CH}_2 - \textbf{CH}_2 - \textbf{CH} - \quad \{Q_\phi(32^{**}) = Q(39)^*\} \\
& \quad\quad\quad\quad\quad\quad | \\
& \quad\quad\quad\quad\quad \textbf{CH}_3 \\
& \quad\quad \textbf{Q3} \\
\text{L:} & - \textbf{CH}_2 - \textbf{CH} - \textbf{CH}_3 \\
& \quad\quad\quad\quad | \\
& \quad\quad\quad\quad \textbf{CH}_3 \quad\quad\quad\quad \phi/\textbf{CH}_4 : \textbf{Q2} = \textbf{Q3} \\
& \quad \textbf{Q2} \\
\text{I:} & - \textbf{CH} - \textbf{CH}_2 - \textbf{CH}_3 \\
& \quad\, | \\
& \quad \textbf{CH}_3
\end{array}
$$

$Q1 = Q(39)/\textbf{CH}_4 -$ protein backbone: $\textbf{Q1} = \textbf{Q4}$

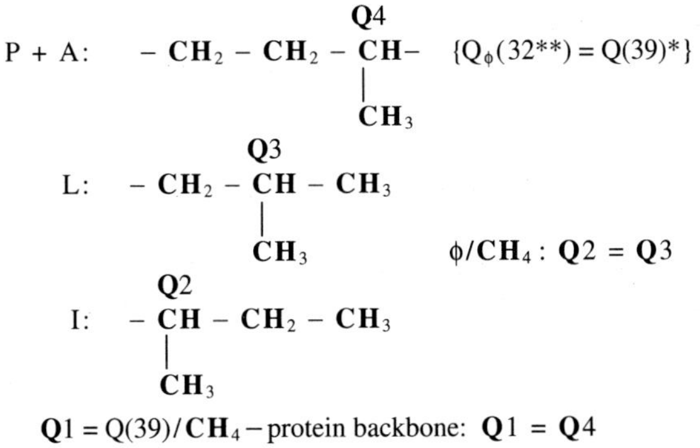

Fig. 4.3 Incident charges **Q1, Q2, Q3, Q4** of segment P + A. **CH$_3$** is the marked point of charge Q(39). **CH$_4$** is the marked point of field ϕ.

Thus, as soon as amino acids I, L have coinciding proton charges and coinciding charge configurations, they can be differentiated as ordinal types. Taking into account that amino acids I, L are the marked points of operator H(43); during DNA translation, methylated DNA bases do not influence the result of translation due to the identity of incident charges **Q2 = Q3**.

We can consider that the identity **Q2 = Q3** for amino acids L, I means scale invariance of amino acids conformal field: amino acid L is a copy of amino acid I independently of the structural differences.

=3=	+ F + G − S	6C	5H	−1O			
=3=	+ F − * − S	6C	4H	−1O			−1*
=3=	+ F − W + Y	5C	2H	1O	−1N		
=3=	+ G − P + R	1C	6H		3N		
=3=	− G − P + Y	4C	1H	1O			
=3=	− * − P + R	1C	5H		3N	−1*	
=3=	+ * − P + Y	4C	2H	1O		1*	
=3=	− P − S + W	5C	4H	−1O	1N		
=3=	+ S + W − Y	3C	8H		1N		

=4=	– A + F – G – ϕ	6C	3H				–1ϕ
		5C	3H				
=4=	– A + F + * – ϕ	6C	4H			1*	–1ϕ
=4=	+ A + G + S + ϕ	2C	7H	1O			1ϕ
		3C	7H	1O			
=4=	+ A – * + S + ϕ	2C	6H	1O		–1*	1ϕ
	+ A – * + **S^{-*}**+ϕ	2C	5H				1S

Homocysteine **$HSCH_2CH_2$–**

=4=	+ A – H + W – ϕ	6C	10H		–1N		–1ϕ
=4=	+ A – R + W + ϕ	6C	5H		–2N		1ϕ
=4=	+ F + G – P + ϕ	4C	3H				1ϕ
		5C	3H				
=4=	+ F – * – P + ϕ	4C	2H			–1*	1ϕ
=4=	+ G + H – S + ϕ	3C	3H	–1O	2N		1ϕ
=4=	– G + P + S – ϕ	4C	7H	1O			–1ϕ
	+ P +S^{-G} – ϕ	2C	6H	1O	1N		
		3C	7H	1O			
=4=	+ G + R – S – ϕ	3C	8H	–1O	3N		–1ϕ
=4=	– G – S – ϕ + Y	6C	3H				–1ϕ
		5C	3H				
=4=	– * + H – S + ϕ	3C	2H	–1O	2N	–1*	1ϕ
=4=	+ * + P + S – ϕ	4C	8H	1O		1*	–1ϕ
	+ S^{P*} – ϕ	2C	6H	1O	1N		
=4=	– * + R – S – ϕ	3C	7H	–1O	3N	–1*	–1ϕ
=4=	+ * – S – ϕ + Y	6C	4H			1*	–1ϕ
=4=	+ H – W + ϕ + Y	2C		1O	1N		1ϕ
		3C		1O	1N		
=4=	+ P + W – ϕ – Y	5C	10H	–1O	1N		–1ϕ
=4=	+ R – W – ϕ + Y	2C	5H	1O	2N		–1ϕ
		1C	5H	1O	2N		

Lysine, Methionine K, M {19^+ 13^-}

=3=	– A + F + G	6C	5H		
=3=	– A + F – *	6C	4H		–1*
=3=	+ A + F – S	7C	7H	–1O	

=3=	+ A – P + R	2C	8H		3N		
=3=	– A – P + W	5C	4H		1N		
	– A + W^{-P}	6C	5H				
=3=	– F + S + W	3C	8H	1O	1N		
=3=	– G + * + H	4C	4H		2N	1*	
=3=	+ G + P + S	4C	9H	1O			
=3=	+ G – S + Y	6C	5H				
=3=	– * + P + S	4C	8H	1O		–1*	
M	+ P^{-*} + **S^{-*}**	3C	7H				1S

Basic configuration of the genetic code.

=3=	– * – S + Y	6C	4H			–1*	
=3=	+ H + R – Y	1C	8H	–1O	5N		
=3=	+ P – R + W	8C	7H		–2N		
=4=	+ A + F – P + ϕ	5C	5H				1ϕ
		6C	5H				
=4=	– A + G + H + ϕ	3C	3H		2N		1ϕ
		4C	3H		2N		
=4=	– A + G + R – ϕ	3C	8H		3N		–1ϕ
		2C	8H		3N		
=4=	– A – G – ϕ + Y	6C	3H	1O			–1ϕ
		5C	3H	1O			
=4=	– A – * + H + ϕ	3C	2H		2N	–1*	1ϕ
=4=	– A – * + R – ϕ	3C	7H		3N	–1*	–1ϕ
=4=	– A + * – ϕ + Y	6C	4H	1O		1*	–1ϕ
=4=	+ A + H – S + ϕ	4C	5H	–1O	2N		1ϕ
=4=	+ A + R – S – ϕ	4C	10H	–1O	3N		–1ϕ
=4=	– A – S + W – ϕ	7C	6H	–1O	1N		–1ϕ
=4=	+ F – G + * – ϕ	7C	6H			1*	–1ϕ
=4=	+ F + H + ϕ – Y	4C	5H	–1O	2N		1ϕ
=4=	– F + P + W – ϕ	5C	10H		1N		–1ϕ
K		4C	10H		1N		

Basic configuration of the genetic code.

=4=	+ F + R – ϕ – Y	4C	10H	–1O	3N		–1ϕ
=4=	+ G – P + ϕ + Y	4C	3H	1O			1ϕ

		5C	3H	1O			
=4=	− * − P + φ + Y	4C	2H	1O		−1*	1φ
=4=	− H + P + R + φ	3C	10H		1N		1φ
K		4C	10H		1N		
=4=	− H + S + W − φ	6C	10H	1O	−1N		−1φ
=4=	− R + S + W + φ	6C	5H	1O	−2N		1φ

Concordance of Amino Acid Conformal Field (CFAA) Metric

Examined CFAA metric tensors can replace amino acids in charge configurations formulae. Each charge configuration induces metric tensor, which can be interpreted as a left and right exchange charge with another charge configuration. Let us list the charge configurations of 0-vector with induced exchange charges (used symbols *p*, *m* are related to the positive and negative parts of the metric tensor):

0-Vector $\{28^{+}\ 6^{-}\}$

=3=	+ F − H − φ	3C	2H		−2N	−1φ	−33*p*	6*m*
=3=	− F + H + φ	−3C	−2H		2N	1φ	33*p*	−6*m*
=3=	+ F − R + φ	3C	−3H		−3N	1φ	27*p*	4*m*
=3=	− F + R − φ	−3C	3H		3N	−1φ	−27*p*	−4*m*
=3=	+ P − S − φ	2C	2H	−1O		−1φ	−30*p*	−2*m*
=3=	− P + S + φ	−2C	−2H	1O		1φ	30*p*	2*m*
=4=	+ A + F − G − Y	1C	2H	−1O			−8*p*	1*m*
=4=	− A − F + G + Y	−1C	−2H	1O			8*p*	−1*m*
=4=	+ A + F + * − Y	1C	3H	−1O		1*	13*p*	22*m*
=4=	− A − F − * + Y	−1C	−3H	1O		−1*	−13*p*	−22*m*
=4=	+ A − F + P + S	−2C	4H	1O			28*p*	11*m*
=4=	− A + F − P − S	2C	−4H	−1O			−28*p*	−11*m*
=4=	+ A − F − S + Y						4*p*	−4*m*
=4=	− A + F + S − Y						−4*p*	4*m*

=4=	+ A − H − P + Y	1C		1O	−2N		1p	4m
=4=	− A + H + P − Y	−1C		−1O	2N		−1p	−4m
=4=	+ F + G + P − W	1C	1H		−1N		26p	13m
=4=	− F − G − P + W	−1C	−1H		1N		−26p	−13m
=4=	+ F − * + P − W	1C			−1N	−1*	5p	−8m
=4=	− F + * − P + W	−1C			1N	1*	−5p	8m
=4=	+ F − H − P + S	1C		1O	−2N		−3p	8m
=4=	− F + H + P − S	−1C		−1O	2N		3p	−8m
=4=	+ F + P − R − S	5C	−1H	−1O	−3N		−3p	2m
=4=	− F − P + R + S	−5C	1H	1O	3N		3p	−2m
=4=	+ G − * + R − Y	−3C	4H	−1O	3N	−1*	10p	−14m
=4=	− G + * − R + Y	3C	−4H	1O	−3N	1*	−10p	14m
=4=	+ G + R + S − W	−4C	2H	1O	2N		29p	11m
=4=	− G − R − S + W	4C	−2H	−1O	−2N		−29p	−11m
=4=	+ G − S + W − Y	1C	3H	−2O	1N		2p	−4m
=4=	− G + S − W + Y	−1C	−3H	2O	−1N		−2p	4m
=4=	+ * − R − S + W	4C	−1H	−1O	−2N	1*	−8p	10m
=4=	− * + R + S − W	−4C	1H	1O	2N	−1*	8p	−10m
=4=	+ * + S - W + Y	−1C	−2H	2O	−1N	1*	19p	25m
=4=	− * − S + W − Y	1C	2H	−2O	1N	−1*	−19p	−25m

The process of CFAA metric concordance can be demonstrated by an example of building a primary instanton as a representation of charge-exchanging group. Primary instanton is built in a three-dimensional Euclidean space by means of binding CFAA charge configurations, which exchange induced charges, into a strand. CFAA charge configurations with charge $Q_p(e)$ with its metric are determined for each exchanging charge e*p*, e*m*. Then one of the possible $Q_p(e)$ charge configurations is selected and the process of metric concordance continues. Process of CFAA metric formally ends by the 0-vector but it is possible to limit the strands of charge-exchanging group by genetic code configuration if to assume the twisting of three-dimensional space. 0-Vector can also be continued. If we designate CFAA metric tensor as $\{\pi^+\pi^-\}$ then charges π^+, π^- are always directed along *z*-axis (Fig. 4.4).

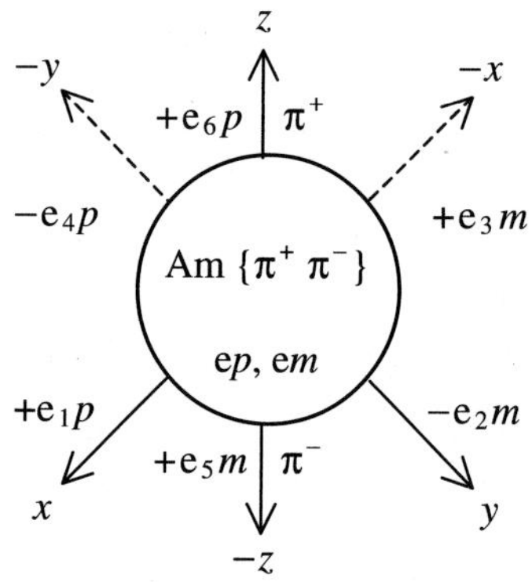

Fig. 4.4 Exchange charges *ep*, *em*, π^+, π^- of amino acid Am. Exchange of charges between charge configurations occurs along coordinates axes *x*, *y*, *z*. For the given direction of axes, charges $+e1p$, $+e3m$, π^+, π^- have positive proper values; charges $-e2m$, $-e4p$ have negative proper values.

Matrix of x, y-axes rotation relatively z-axis looks as:

$$Zxy = \begin{vmatrix} e_4p & e_3m \\ e_1p & e_2m \end{vmatrix}$$

Table 4.1 contains metric tensors of charges $Q_p(e)$ $\{\pi^+ \pi^-\}$; charges of the genetic code amino acids are highlighted.

For each charge $Q_p(e)$, it is necessary to select two charge configurations as minimum. In a simplest case, charge configurations are selected in such a way that one exchanging charge is located along each coordinate axis; in more complex cases, it is possible to use superposition of the exchanging charges.

If we take into account the self-acting of charges $Q_p(e)$, then each charge configuration determines those amino acids, which undergo the process of the charge exchange. In this case, it is necessary to select only the charge configurations of those charges $Q_p(e)$, which are equal to the charges of the genetic code amino acids. For example, amino acid Y $\{12^+ 7^-\}$ with charge configuration A + F – G: **8C9H** $+4p$ $+8m$ can also exchange **metric charges of amino acids** A $\{12^+ 5^-\}$, F $\{8^+ 8^-\}$, G $\{-16^+ -5^-\}$ on three coordinate axes with six charge configurations and determines the distance between charge configurations in **units of radical length** of corresponding amino acid (Fig. 4.5).

The distance between two charge configurations depends on two amino acids and in general case is different for two time points.

Table 4.1 Metric tensor $\{\pi^+ \pi^-\}$ of proton charge $Q_p(e)$ of amino acids conformal field

Protonic charge $Q_p(e)$	*Metric tensor* $\{\pi^+ \pi^-\}$	*Protonic charge* $Q_p(e)$	*Metric tensor* $\{\pi^+ \pi^-\}$	*Protonic charge* $Q_p(e)$	*Metric tensor* $\{\pi^+ \pi^-\}$
*	$\{16^+ 5^-\}$	29	$\{18^+ 7^-\}$	59	$\{17^+ 7^-\}$
0	$\{28^+ 6^-\}$	30	$\{30^+ 3^-\}$	60	$\{20^+ 5^-\}$
G(1)	$\{16^+ 5^-\}$	D,N(31)	$\{19^+ 13^-\}$	61	$\{15^+ 5^-\}$
2	$\{31^+ 5^-\}$	32	$\{35^+ 4^-\}$	62	$\{26^+ 5^-\}$
3	$\{18^+ 8^-\}$	I,L(33)	$\{19^+ 14^-\}$	63	$\{13^+ 9^-\}$
4	$\{30^+ 4^-\}$	34	$\{33^+ 3^-\}$	64	$\{30^+ 3^-\}$
5	$\{14^+ 9^-\}$	35	$\{20^+ 9^-\}$	65	$\{14^+ 13^-\}$
6	$\{30^+ 0^-\}$	36	$\{25^+ 3^-\}$	66	$\{25^+ 5^-\}$
7	$\{13^+ 16^-\}$	37	$\{17^+ 5^-\}$	67	$\{13^+ 9^-\}$
8	$\{36^+ 4^-\}$	38	$\{27^+ 5^-\}$	68	$\{23^+ 3^-\}$
A(9)	$\{12^+ 5^-\}$	E,Q(39)	$\{18^+ 11^-\}$	69	$\{17^+ 4^-\}$
10	$\{32^+ 3^-\}$	40	$\{33^+ 3^-\}$	70	$\{23^+ 2^-\}$
11	$\{21^+ 8^-\}$	K,M(41)	$\{19^+ 13^-\}$	71	$\{16^+ 7^-\}$
12	$\{24^+ 6^-\}$	42	$\{40^+ 3^-\}$	72	$\{17^+ 4^-\}$
13	$\{23^+ 7^-\}$	H(43)	$\{11^+ 2^-\}$	W(73)	$\{16^+ 8^-\}$
14	$\{29^+ 5^-\}$	44	$\{16^+ 4^-\}$	74	$\{21^+ 1^-\}$
15	$\{16^+ 14^-\}$	45	$\{20^+ 4^-\}$	75	$\{14^+ 8^-\}$
16	$\{40^+ 4^-\}$	46	$\{24^+ 4^-\}$	76	$\{21^+ 1^-\}$
S(17)	$\{12^+ 8^-\}$	47	$\{17^+ 10^-\}$	77	$\{14^+ 3^-\}$
18	$\{28^+ 4^-\}$	48	$\{31^+ 4^-\}$	78	$\{17^+ 4^-\}$
19	$\{21^+ 7^-\}$	F(49)	$\{8^+ 8^-\}$	79	$\{9^+ 7^-\}$
20	$\{22^+ 5^-\}$	50	$\{30^+ 7^-\}$	80	$\{28^+ 3^-\}$
21	$\{29^+ 6^-\}$	51	$\{13^+ 9^-\}$	81	$\{12^+ 10^-\}$
22	$\{28^+ 3^-\}$	52	$\{22^+ 4^-\}$	82	$\{29^+ 0^-\}$
P(23)	$\{12^+ 6^-\}$	53	$\{17^+ 4^-\}$	83	$\{10^+ 7^-\}$
24	$\{37^+ 4^-\}$	54	$\{14^+ 4^-\}$	84	$\{21^+ 2^-\}$
C,T,V(25)	$\{20^+ 14^-\}$	R(55)	$\{11^+ 4^-\}$	-	-
26	$\{30^+ 4^-\}$	56	$\{18^+ 3^-\}$	200	$\{1^+ 0^-\}$
27	$\{20^+ 8^-\}$	Y(57)	$\{12^+ 7^-\}$	202	$\{2^+ 0^-\}$
28	$\{28^+ 4^-\}$	58	$\{24^+ 3^-\}$	204	$\{1^+ 0^-\}$

Let us select an additional charge configuration for amino acid Y

=3=   – * – S + W   8**C**   9**H**   –1**O**   1**N**   –1*   –7*p* –18*m*

with metric charges determined by amino acids $-*\{5^+ 16^-\}$, $S\{-12^+ -8^-\}$, $W\{16^+ 8^-\}$.

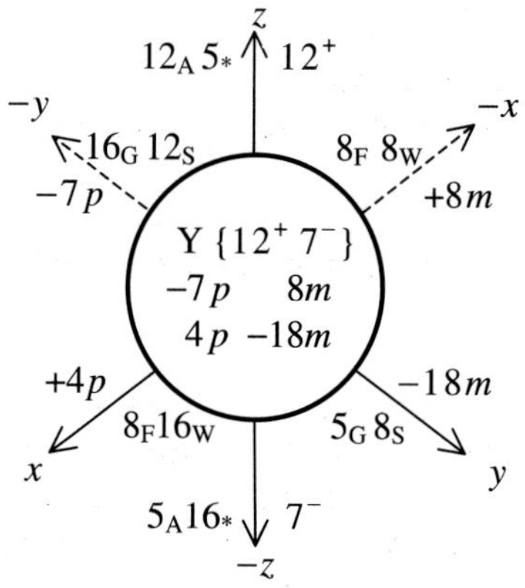

Fig. 4.5 Exchanging charges of amino acid Y. Distances to the charge configurations along coordinate axes *x*, *y*, *z* are determined in units of amino acid radical lengths F, G, A, *, S, W for two time points.

Basic length of the terminator can be selected equal to the length of amino acid –G, i.e., to the length of protonic bond. Placing of radicals lengths of amino acids A, F, *, W on coordinate axes determines dynamic state of the site Y. Obtained model of instanton site determines an amino acid conformal field with metric $4^{+}2^{-}$.

If we select the additional basic configuration of amino acid Y as

$$=4= \quad -A-G+W-\phi \quad 8C \quad 8H \quad 1N \quad -1\phi \ -48p \ -4m,$$

then, the presence of the field $+1\phi$ changes position of charges ep, em relatively to coordinate axes but does not change the CFAA metric, while presence of the field -1ϕ not only changes position of charges ep, em but also introduces CFAA metric $3^{+}3^{-}$. It is possible to select a position of ep, em in the following way:

$$+1\phi: 4^{+}2^{-}: \mathrm{Am}(ep, em); -\mathrm{Am}(em, ep)$$
$$-1\phi: 3^{+}3^{-}: \mathrm{Am}(-ep, em); -\mathrm{Am}(em, -ep).$$

Then 4_{ϕ}-index configurations are different from three-index configurations by involutive transformations of coordinate axes with preservation or change of CFAA metric.

Matrix of turn of x, y-axes relatively to z-axis for amino acid Y looks as:

$$\mathrm{Y}z = \begin{vmatrix} -7p & +8m \\ +4p & -18m \end{vmatrix}$$

It is possible to select the following configurations for exchange charges $Q_p(e)$:

$Q_p(20^*)$	$+A-H+R-S$		5H	1O*	1N		$0p$ $-1m$
$Q_p(7)$	$-F-G+Y$		-1H	1O			$-12p$ $-6m$
$Q_p(7)$	$+F+G-H$	3C	3H		-2N		$13p$ $11m$
$Q_p(8)$	$-A+P-\phi$	2C	2H			-1ϕ	$-30p$ $1m$
$Q_p(12)$	$-F+H-R+W$	2C					$2p$ $-4m$
$Q_p(18)$	$-H+P+R-S$	2C	7H	-1O	1N		$0p$ $0m$

Z-axis for the charges $-7p, -18m, 7^-$ is directed oppositely to z-axis of amino acid Y. For the charge $+4p$, the marked point **O*** of external hydrogenous bond is introduced. This marked point increases the charge $+4p$ up to $+20^*p$ and simultaneously changes the z-axis orientation to the opposite to z-axis of amino acid Y.

It is necessary to take into account the metric of field $\phi\{30^+\ 0^-\}$: $30_\phi = 30_C$; $0_\phi = 0$ and permutation of charges ep, em when building instanton site $Q_p(8)$.

When building instanton site $Q_p(12)$, it is necessary to select a basic amino acid (from four), which will be the instanton site, for example, H. This operation can be applied also for the other (including three-index) configurations if charges $Q_p(e)$ do not coincide with the charges of basic amino acids of the genetic code. Then charge $Q_p(12)$ $\{24^+\ 6^-\}$ transforms into $Q_p(H)$ $\{11^+\ 2^-\}$:

$Q_p(H)$	$-F-R+W$	2C			$2p$ $-4m$	$\{24^+\ 6^-\}$
$Q_p(H)$	$+A-P+Y$	5C	5H	1O	$12p$ $6m$	$\{11^+\ 2^-\}$

and three-index configuration of amino acid H is introduced. Both configurations determine instanton site with the **dominant amino acid** H over charge $Q_p(12)$.

If we determine the dominant amino acid Am′, which is located in the representation of amino acid Am, for each instanton site of basic configuration of genetic code amino acid Am, then two-charge pair Q_p(Am), Q_p(Am′) completely determines the conformal field of amino acids and is a bundle of ordinal type ω_2.

Each bond of instanton sites α, β along y-axis determines the connection matrix in units of radical lengths of four amino acids $\alpha 1, \alpha 2, \beta 1, \beta 2$:

$$\Gamma_{\alpha\beta}{}^{\varphi} = \begin{vmatrix} e\beta 1 & e\beta 2 \\ e\alpha 1 & e\alpha 2 \end{vmatrix},$$

which is the matrix of plane $y_\alpha\, y_\beta$ twisting (Fig. 4.6).

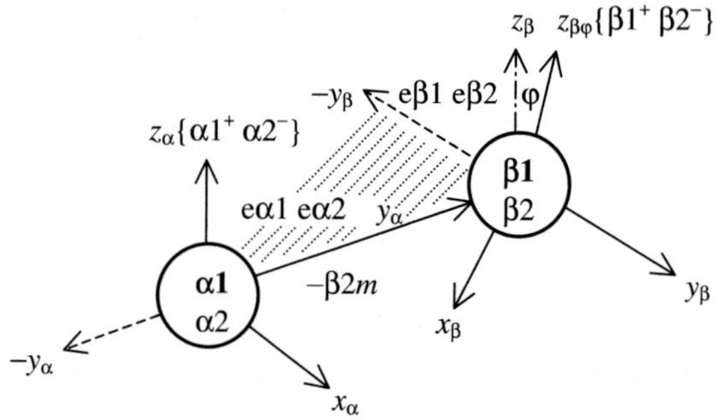

Fig. 4.6 Two-charge bundle of amino acids $\alpha 1$, $\alpha 2$, $\beta 1$, $\beta 2$ in instanton junctions α, β. Twisting angle φ of the triangular plane $y_\alpha\ y_\beta$ transfers z-axis into z_φ of angle β. Dominant amino acids $\alpha 1$, $\beta 1$ are contained in the linear representation of basic amino acids $\alpha 2$, $\beta 2$. Charges $z_\alpha\{\alpha 1^+\ \alpha 2^-\}$ determine the connection form by z-axis.

When binding instanton sites along z-axis, it is necessary to concord the twisting matrix $\Gamma_{\alpha\gamma}{}^{\varphi z}$ with matrix of connection form $z_\gamma\, z_\alpha$, determined by metric coefficients of axes $z_\gamma\{\gamma 1^-\ \gamma 2^+\}$ and $z_\alpha\{\alpha 1^+\ \alpha 2^-\}$:

$$\Gamma_{\alpha\gamma}{}^{\varphi z} = \begin{vmatrix} \gamma 1^- e\gamma 1 & \gamma 2^+ e\gamma 2 \\ \alpha 1^+ e\alpha 1 & \alpha 2^- e\alpha 2 \end{vmatrix}$$

Thus, in order to build a primary instanton, it is necessary to determine two charge configurations in each connection site by one of the examined modes and to determine dominant and basic charge of CFAA bundle.

Examined charge configurations of amino acids clearly take into account the basic property of CFAA states: each state of the amino acid conformal field is determined by quantum numbers ±Am, which actually are the amino acids themselves. Therefore, representations of the amino acids contain only coefficients ±1.

Auto-dual CFAA Bundle

Each charge configuration of the charge $Q_p(e)$ can be replaced by its three- or four-index representation and transformed into the recursive linear strand of amino acids. 4_ϕ-index configurations can be transformed in two ways: either by substitution of evident four-index representation or by branching with the basic amino acid Am and factorization of the field ϕ. For example:

$$
\begin{array}{c}
-G \\
+ \\
Y \rightarrow -\underline{A} + H + \underline{P} \rightarrow -\underline{F - S + Y} + R - \underline{F + * + W} \rightarrow \cdots \\
+ \\
-S
\end{array}
$$

20-Parametric Lie Group of CFAA Duplication

All amino acids with isolated charges have different metric CFAA tensors, whereas metric tensors $19^{+}\,13^{-}$ of amino acids with coinciding charges (D, N) and (K, M) also coincide: accordingly to the genetic code

$$Q_p(2D + 2N) = Q_p(2K + M - *) = \mathbf{GF}(5^{3*}) = \mathbf{GF}(41) + \mathbf{GF}(83)$$

and CFAA duplication is determined by the equation

$$2Q_p(K - N) = Q_p(2D - M - *) = 20\text{-parametric Lie group.}$$

Let us introduce the bundle of amino acids into the field **GF**(83). Then each amino acid is determined by charge **GF**(Am) and variation of field $\delta\phi$. Basic configuration of amino acid **GF**(Am) and variation of field $\delta\phi$ are selected in such a way that there was an identity: Am = **GF**(Am) + $\delta\phi$ (Table 4.2).

All amino acids in field **GF**(83) have isolated charges and, therefore, they can be represented by the linear combinations of remained 19 amino acids, terminator and binding field ϕ.

Table 4.2 Bundle of amino acids of the genetic code in field **GF**(83)

Amino acid Am	*Protonic charge **GF**(Am)*	*Basic configuration*				*Field variation $\delta\phi$*
*	− 1		−1H			0
G	+ 1		+1H			0
A	+ 3	1C	4H		−1N	−1H +1N
S	+ 4	1C	4H	1O	−2N	−1H +2N
P	+ 7	3C	5H		−1S	+1S
C	+ 9	1C	3H			+1S
T	+ 11	2C	5H	1O	−2N	+2N
V	+ 16	3C	6H	-1O		+1H +1O
D	+ 17	2C	3H	2O	−2N	+2N
N	+ 23	2C	4H		1N	+1O

Contd.

Table 4.2 Contd

I	+ 25	**4C**	**9H**	-1**O**		+1**O**
L	+ 27	**3C**	**9H**			+1**C**
Q	+ 29	**3C**	**4H**		1**N**	+2**H** +1**O**
E	+ 31	**3C**	**5H**	1**O**		+1**O**
K	+ 37	**3C**	12**H**		1**N**	−2**H** +1**C**
M	+ 41	**3C**	**7H**		1**S**	0
H	+ 49	**5C**	**5H**	2**N**		−1**C**
F	+ 59	**7C**	**9H**	1**O**		−2**H** −1**O**
R	+ 61	**5C**	10**H**		3**N**	−1**C**
Y	+ 64	**7C**	**7H**	1**O**	1**N**	−1**N**
W	+ 81	**9C**	12**H**	1**O**	1**N**	−1**O**

Amino acids **E, Q** are marked points of field **GF**(83), which are transposed instead of infinitely-remote point I, L.

Identity of charge configurations of amino acids I, L determines CFAA self-acting, but in the field **GF**(83), these configurations form CFAA metric tensor.

Table 4.3 shows metric tensors of charges Q_p(e) in field GF(83). These charges are more than protonic charges of amino acids as well as protonic charges of atoms. Therefore, conformal field of amino acids has a molecular basis.

Table 4.3 Metric tensor $\{\pi^+ \pi^-\}$ of protonic charge Q_pGF(83) of amino acids

Protonic charge Q_p***GF83****(e)*	*Metric tensor* $\{\pi^+ \pi^-\}$	*Protonic charge* Q_p***GF83****(e)*	*Metric tensor* $\{\pi^+ \pi^-\}$	*Protonic charge* Q_p***GF83****(e)*	*Metric tensor* $\{\pi^+ \pi^-\}$
*	$\{477^+\ 73^-\}$	Q(29)	$\{463^+\ 74^-\}$	**F**(59)	$\{371^+\ 58^-\}$
0	$\{706^+\ 54^-\}$	30	$\{640^+\ 52^-\}$	60	$\{500^+\ 33^-\}$
G(1)	$\{477^+\ 73^-\}$	E,D,N(31)	$\{447^+\ 75^-\}$	**R**(61)	$\{357^+\ 55\text{-}\}$
2	$\{670^+\ 60^-\}$	32	$\{608^+\ 58^-\}$	62	$\{513^+\ 38^-\}$
A(3)	$\{477^+\ 76^-\}$	I,L(33)	$\{515^+\ 87^-\}$	63	$\{412^+\ 57^-\}$
S(4)	$\{593^+\ 50^-\}$	34	$\{595^+\ 63^-\}$	**Y**(64)	$\{422^+\ 32^-\}$
5	$\{571^+\ 91^-\}$	35	$\{549^+\ 80^-\}$	65	$\{374^+\ 53^-\}$
6	$\{588^+\ 49^-\}$	36	$\{602^+\ 61^-\}$	66	$\{460^+\ 41^-\}$
P(7)	$\{469^+\ 75^-\}$	**K**(37)	$\{454^+\ 66^-\}$	67	$\{394^+\ 56^-\}$
8	$\{678^+\ 63^-\}$	38	$\{612^+\ 52^-\}$	68	$\{469^+\ 32^-\}$
C,A(9)	$\{473^+\ 81^-\}$	E,Q(39)	$\{505^+\ 88^-\}$	69	$\{383^+\ 63^-\}$
10	$\{675^+\ 54^-\}$	40	$\{615^+\ 49^-\}$	70	$\{465^+\ 28^-\}$
T(11)	$\{475^+\ 82^-\}$	K,**M**(41)	$\{416^+\ 66^-\}$	71	$\{365^+\ 59^-\}$

Contd.

Table 4.3 Contd

12	$\{672^+\ 60^-\}$	42	$\{607^+\ 51^-\}$	72	$\{439^+\ 32^-\}$
13	$\{581^+\ 94^-\}$	H(43)	$\{503^+\ 74^-\}$	W(73)	$\{384^+\ 47^-\}$
14	$\{668^+\ 71^-\}$	44	$\{584^+\ 59^-\}$	74	$\{428^+\ 38^-\}$
15	$\{585^+\ 86^-\}$	45	$\{510^+\ 66^-\}$	75	$\{354^+\ 43^-\}$
V(16)	$\{585^+\ 47^-\}$	46	$\{603^+\ 53^-\}$	76	$\{427^+\ 34^-\}$
D,S(17)	$\{470^+\ 70^-\}$	47	$\{484^+\ 75^-\}$	77	$\{336^+\ 46^-\}$
18	$\{668^+\ 59^-\}$	48	$\{561^+\ 45^-\}$	78	$\{413^+\ 24^-\}$
19	$\{556^+\ 101^-\}$	**H**,F(49)	$\{417^+\ 66^-\}$	79	$\{325^+\ 51^-\}$
20	$\{665^+\ 56^-\}$	50	$\{574^+\ 37^-\}$	80	$\{411^+\ 22^-\}$
21	$\{542^+\ 103^-\}$	51	$\{482^+\ 80^-\}$	**W**(81)	$\{259^+\ 33^-\}$
22	$\{660^+\ 59^-\}$	52	$\{538^+\ 43^-\}$	82	$\{389^+\ 27^-\}$
N,P(23)	$\{467^+\ 73^-\}$	53	$\{453^+\ 67^-\}$	83	$\{338^+\ 42^-\}$
24	$\{608^+\ 67^-\}$	54	$\{522^+\ 53^-\}$	84	$\{355^+\ 30^-\}$
I,C,T,V(25)	$\{510^+\ 64^-\}$	R(55)	$\{457^+\ 63^-\}$	-	
26	$\{617^+\ 67^-\}$	56	$\{518^+\ 48^-\}$	235	$\{3^+\ 0^-\}$
L(27)	$\{463^+\ 74^-\}$	Y(57)	$\{434^+\ 69^-\}$	237	$\{1^+\ 0^-\}$
28	$\{650^+\ 59^-\}$	58	$\{497^+\ 37^-\}$	238	$\{1^+\ 0^-\}$

Q(29), E(31) – marked involutive pair of field **GF**(83)
Basic amino acids in field **GF**(83) are allocated by a bold font.

We can deduce from Table 4.3 that the conformal field of amino acids has been duplicated and has metric 13^+7^- in the mixed gauge:

$$G(1), \mathbf{G}(1), \mathbf{A}(3), \mathbf{S}(4), \mathbf{P}(7), \mathbf{T}(11), \mathbf{V}(16), \mathbf{L}(27), \underline{Q}(29), \mathbf{K}(37),$$
$$M(41), \mathbf{M}(41), H(43), R(55), Y(57), \mathbf{F}(59), \mathbf{R}(61), \mathbf{Y}(64), W(73), \mathbf{W}(81).$$

Marked point of terminators has been transferred from amino acid P to amino acid Q.

The identity of DNA transcription is true in the field **GF**(83):

$$Q_4\{C(9) - T(11) + D(17) - N(23) + I(25) - L(27) + \underline{Q}(29) - E(31) + K(37) + M(41) + V(16^*)\} =$$
$$Q_4\{[C(9) + D(17) + \underline{Q}(29)] + [I(25) + K(37) + V(16^*)] - [L(27) + E(31)] + [M(41) - T(11) - N(23)]\} =$$
$$Q_p\{[T55 + \mathbf{G}78 - \mathbf{C}58]_{PHK} + 8^*\} =$$
$$Q_p\{[\mathbf{T}66]_{PHK} + V(16)\} = \mathbf{GF}(83^*).$$

Informational nucleotide **T** has been transferred into connection, V(16) is the production of entropy.

Thus, during the amino acids bundling in the field **GF**(83), self-acting of the field **GF**(83) occurs: the amino acids charges are paired with formation of two-point Riemann space. Each amino acid is characterized by two CFAA charges in field **GF**(83): basic **GF**(Am) and dominant Am and corresponding configurations.

Metric M{416^+ 66^-} in field **GF**(83) is in concordance with connection because for complementary DNA pairs the following identity is true:

$$\mathbf{GF}(3) \times \mathbf{GF}(137) + \overbrace{\mathbf{GF}(3^*) + \mathbf{GF}(3)}^{\phi^*} = 416; \mathbf{T} = Q_p(66).$$

Fields **GF**(3) and **GF**(3*) $\to$ **GF**(2) are the fields of parallel transfer of informational nucleotides of DNA and RNA correspondingly, nucleotide T is the connecting link for the basic and informational DNA and RNA nucleotides. Index p shows that bound proton, which is equivalent to antiproton, is the generator of fields **GF**(3), **GF**(2).

Field **GF**(83) performs selection of dominant and basic amino acids configurations. This situation differs from the one, which was described for 10-parametrical Lie group. But for exchanging charges as well as for charges, which do not coincide with amino acids charges in the field **GF**(83), it is again necessary to select the dominant configuration and the dominant amino acid, which is contained in representation of exchanging charge; in this case, basic charge of amino acid is proposed by field **GF**(83) itself.

With further expansion of field ϕ, exchanging charges of various chemical compounds, including DNA and RNA, can be expressed through equivalent charge configurations of amino acids.

Let us make an extract of dominant configurations of the genetic code amino acids from general totality of configurations for the charge Q_p(Am) in the field **GF**(83). Table 4.4 contains configurations of the genetic code amino acids (the basic charge in the field **GF**(83) is associated with pseudo-symbol ζ).

Dominant configurations of the genetic code in field GF(83)

Table 4.4

Terminator* {6^+ 4^-}				
=3=	– I + L + *		1*	1ζ
=3=	– A – V + I	–1H		6ζ
=3=	– A – V + L	–1H		8ζ
=3=	– A – D + E	–1H		11ζ

Contd.

Table 4.4 Contd

=4=	+ G + S − T + ϕ	−1**C**	−1**H**		1ϕ	0ζ
=4=	+ S − T − * + ϕ	−1**C**	−2**H**	−1*	1ϕ	0ζ
=4=	+ N − Q − * + ϕ	−1**C**	−2**H**	−1*	1ϕ	1ζ
=4=	+ G − A − * + ϕ	−1**C**	−2**H**	−1*	1ϕ	5ζ
=4=	− A − P + V + ϕ	−1**C**	−1**H**		1ϕ	12ζ
=4=	− P − C + M + ϕ	−1**C**	−1**H**		1ϕ	31ζ
0–Vector $\{27^+\ 0^-\}$						
=4=	+ S − T − N + Q					−1ζ
=4=	− S + T + N − Q					1ζ
=4=	+ G − I + L + *			1*		2ζ
=4=	− G − I + L − *			−1*		2ζ
=4=	+ S − T − V + I					2ζ
=4=	+ V − D − L + E					3ζ
=4=	− V + N + I − Q					3ζ
=4=	+ G − A − N + Q					4ζ
=4=	− A − N + Q − *		−1**H**	−1*		4ζ
=4=	+ G − A − S + T					5ζ
=4=	− A − S + T − *		−1**H**	−1*		5ζ
=4=	+ V − D − I + E					5ζ
=4=	− V + N + L − Q					5ζ
=4=	+ G − A − V + I					7ζ
=4=	− A − V + I − *		−1**H**	−1*		7ζ
=4=	+ S − T − D + E					7ζ
=4=	− D + N − Q + E					8ζ
=4=	+ G − A − V + L					9ζ
=4=	− A − V + L − *		−1**H**	−1*		9ζ
=4=	− T + K + Y − W					9ζ
=4=	+ G − A − D + E					12ζ
=4=	− A − D + E − *		−1**H**	−1*		12ζ
=4=	− G − A − P + I					14ζ
=4=	− A − P + I + *		1**H**	1*		14ζ
=4=	− G − A − P + L					16ζ
=4=	− A − P + L + *		1**H**	1*		16ζ
=4=	+ A − C − V + M					19ζ
Glycine G $\{10^+\ 7^-\}$						
=3=	+ G + I − L		1**H**			−1ζ
=3=	+ G − I + L		1**H**			3ζ
=3=	− I + L − *			−1*		3ζ

Contd.

Table 4.4 Contd

=3=	−G−P+V		1H			8ζ
=3=	−P+V+*		2H	1*		8ζ
=3=	−A−P+I		1H			15ζ
=3=	−A−P+L		1H			17ζ
=4=	−G−N+Q−ϕ	1C	1H		−1ϕ	−1ζ
=4=	−G−S+T−ϕ	1C	1H		−1ϕ	0ζ
=4=	−S+T+*−ϕ	1C	2H	1*	−1ϕ	0ζ
=4=	−G−V+I−ϕ	1C	1H		−1ϕ	2ζ
=4=	−V+I+*−ϕ	1C	2H	1*	−1ϕ	2ζ
=4=	−G−V+L−ϕ	1C	1H		−1ϕ	4ζ
=4=	−V+L+*−ϕ	1C	2H	1*	−1ϕ	4ζ
=4=	−G−D+E−ϕ	1C	1H		−1ϕ	7ζ
=4=	−D+E+*−ϕ	1C	2H	1*	−1ϕ	7ζ
=4=	−A−C+M−ϕ	1C	1H		−1ϕ	23ζ
Field Basis ϕ {19⁺ 2⁻}						
=3=	+I−L+ϕ				1ϕ	4ζ
=3=	−I+L+ϕ				1ϕ	8ζ
=4=	+P−T−F+Y	1C				1ζ
=4=	−G−N+Q+*	1C	1H	1*		4ζ
=4=	−G−S+T+*	1C	1H	1*		5ζ
=4=	+P−V−D+E	1C				5ζ
=4=	−T+I+Q−K	1C				6ζ
=4=	−G−V+I+*	1C	1H	1*		7ζ
=4=	−S+N+I−K	1C				7ζ
=4=	−S+D+F−Y	1C				8ζ
=4=	−T+L+Q−K	1C				8ζ
=4=	−G−V+L+*	1C	1H	1*		9ζ
=4=	−S+N+L−K	1C				9ζ
=4=	−G−D+E+*	1C	1H	1*		12ζ
=4=	+P−C−L+M	1C				12ζ
=4=	+P−C−I+M	1C				14ζ
=4=	−T+E+F−Y	1C				15ζ
=4=	−P+K+H−R	1C				18ζ
=4=	+A−S−T+E	1C				19ζ
=4=	−G−A−C+M	1C				28ζ
=4=	−A−C+M+*	1C	1H	1*		28ζ
Alanine A {3⁺ 17⁻}						
=3=	+S+F−Y	1C	3H			−1ζ
=3=	+A+I−L	1C	3H			1ζ
=3=	+A−I+L	1C	3H			5ζ

Contd.

Table 4.4 Contd

=3=	+ G − N + Q	1**C**	3**H**				7ζ
=3=	− N + Q − *	1**C**	2**H**		−1*		7ζ
=3=	+ G − S + T	1**C**	3**H**				8ζ
=3=	− S + T − *	1**C**	2**H**		−1*		8ζ
=3=	+ G − V + I	1**C**	3**H**				10ζ
=3=	− V + I − *	1**C**	2**H**		−1*		10ζ
=3=	+ G − V + L	1**C**	3**H**				12ζ
=3=	− V + L − *	1**C**	2**H**		−1*		12ζ
=3=	+ G − D + E	1**C**	3**H**				15ζ
=3=	− D + E − *	1**C**	2**H**		−1*		15ζ
=3=	− G − P + I	1**C**	3**H**				17ζ
=3=	− P + I + *	1**C**	4**H**		1*		17ζ
=3=	− G − P + L	1**C**	3**H**				19ζ
=3=	− P + L + *	1**C**	4**H**		1*		19ζ
=4=	+ G − A + P − ϕ	2**C**	3**H**			−1ϕ	−1ζ
=4=	− G − C + M − ϕ	2**C**	3**H**			−1ϕ	25ζ
=4=	− C + M + * − ϕ	2**C**	4**H**		1*	−1ϕ	25ζ
Serine S $\{9^+ 8^-\}$							
=3=	+ T + V − L	1**C**	3**H**	1**O**			0ζ
=3=	+ S + I − L	1**C**	3**H**	1**O**			2ζ
=3=	+ T + V − I	1**C**	3**H**	1**O**			2ζ
=3=	+ T + N − Q	1**C**	3**H**	1**O**			5ζ
=3=	+ S − I + L	1**C**	3**H**	1**O**			6ζ
=3=	+ A − F + Y	1**C**	3**H**	1**O**			8ζ
=3=	+ G − A + T	1**C**	3**H**	1**O**			9ζ
=3=	− A + T − *	1**C**	2**H**	1**O**	−1*		9ζ
=4=	+ V + Q − K − ϕ	2**C**	3**H**	1**O**		−1ϕ	2ζ
=4=	− G + T + * − ϕ	2**C**	4**H**	1**O**	1*	−1ϕ	3ζ
=4=	+ N + I − K − ϕ	2**C**	3**H**	1**O**		−1ϕ	5ζ
=4=	+ N + L − K − ϕ	2**C**	3**H**	1**O**		−1ϕ	7ζ
=4=	+ A − S + D − ϕ	2**C**	3**H**	1**O**		−1ϕ	10ζ
=4=	+ A − P + T + ϕ	3**H**	1**O**			1ϕ	13ζ
=4=	+ A − T + E − ϕ	2**C**	3**H**	1**O**		−1ϕ	17ζ
=4=	+ G − S + E − ϕ	2**C**	3**H**	1**O**		−1ϕ	22ζ
=4=	− S + E − * − ϕ	2**C**	2**H**	1**O**	−1*	−1ϕ	22ζ
Proline P $\{16^+ 7^-\}$							
=3=	+ P + I − L	3**C**	5**H**				5ζ
=3=	+ P − I + L	3**C**	5**H**				9ζ
=3=	− G + V + *	3**C**	6**H**		1*		14ζ

Contd.

Table 4.4 Contd

=3=	− G − A + I	3**C**	5**H**					21ζ
=3=	− A + I + *	3**C**	6**H**			1*		21ζ
=3=	− G − A + L	3**C**	5**H**					23ζ
=3=	− A + L + *	3**C**	6**H**			1*		23ζ
=4=	+ C + I − M + ϕ	2**C**	5**H**				1ϕ	−1ζ
=4=	+ C + L − M + ϕ	2**C**	5**H**				1ϕ	1ζ
=4=	+ V + D − E + ϕ	2**C**	5**H**				1ϕ	8ζ
=4=	+ T + F − Y + ϕ	2**C**	5**H**				1ϕ	12ζ
=4=	+ A − N + Q + ϕ	2**C**	5**H**				1ϕ	15ζ
=4=	+ S − T + V + ϕ	2**C**	5**H**				1ϕ	15ζ
=4=	+ A − S + T + ϕ	2**C**	5**H**				1ϕ	16ζ
=4=	+ A − V + I + ϕ	2**C**	5**H**				1ϕ	18ζ
=4=	+ K + H − R − ϕ	4**C**	5**H**				−1ϕ	19ζ
=4=	+ G − A + V + ϕ	2**C**	5**H**				1ϕ	20ζ
=4=	+ A − V + L + ϕ	2**C**	5**H**				1ϕ	20ζ
=4=	− A + V − * + ϕ	2**C**	4**H**			−1*	1ϕ	20ζ
=4=	+ K + F − W + ϕ	2**C**	5**H**				1ϕ	21ζ
=4=	+ A − D + E + ϕ	2**C**	5**H**				1ϕ	23ζ
=4=	+ G − C + M + ϕ	2**C**	5**H**				1ϕ	39ζ
=4=	− C + M − * + ϕ	2**C**	4**H**			−1*	1ϕ	39ζ
Cysteine C $\{4^+ 3^-\}$								
=3=	+ C + I − L	1**C**	3**H**		1**S**			7ζ
=3=	+ C − I + L	1**C**	3**H**		1**S**			11ζ
=3=	+ A − V + M	1**C**	3**H**		1**S**			28ζ
=4=	+ P − L + M − ϕ	2**C**	3**H**		1**S**		−1ϕ	15ζ
=4=	+ P − I + M − ϕ	2**C**	3**H**		1**S**		−1ϕ	17ζ
=4=	− G − A + M − ϕ	2**C**	3**H**		1**S**		−1ϕ	31ζ
=4=	− A + M + * − ϕ	2**C**	4**H**		1**S**	1*	−1ϕ	31ζ
Threonine T $\{7^+ 9^-\}$								
=3=	− G + A + S	2**C**	5**H**	1**O**				6ζ
=3=	+ A + S + *	2**C**	6**H**	1**O**		1*		6ζ
=3=	+ T + I − L	2**C**	5**H**	1**O**				9ζ
=3=	+ S − N + Q	2**C**	5**H**	1**O**				10ζ
=3=	+ S − V + I	2**C**	5**H**	1**O**				13ζ
=3=	+ T − I + L	2**C**	5**H**	1**O**				13ζ
=3=	+ S − V + L	2**C**	5**H**	1**O**				15ζ
=3=	+ S − D + E	2**C**	5**H**	1**O**				18ζ
=3=	+ K + Y − W	2**C**	5**H**	1**O**				20ζ
=4=	− A + S + P − ϕ	3**C**	5**H**	1**O**			−1ϕ	2ζ

Contd.

Table 4.4 Contd

=4=	+ I + Q − K − ϕ	3**C**	5**H**	1**O**		−1ϕ	11ζ
=4=	+ G + S − * + ϕ	1**C**	4**H**	1**O**	−1*	1ϕ	12ζ
=4=	+ L + Q − K − ϕ	3**C**	5**H**	1**O**		−1ϕ	13ζ
=4=	+ S − P + V + ϕ	1**C**	5**H**	1**O**		1ϕ	19ζ
=4=	+ E + F − Y − ϕ	3**C**	5**H**	1**O**		−1ϕ	20ζ
=4=	+ A − S + E − ϕ	3**C**	5**H**	1**O**		−1ϕ	24ζ
Valine V {9⁺ 14⁻}							
=3=	+ G + P − *	3**C**	6**H**		−1*		9ζ
=3=	+ D + I − E	3**C**	7**H**				11ζ
=3=	+ D + L − E	3**C**	7**H**				13ζ
=3=	+ V + I − L	3**C**	7**H**				14ζ
=3=	+ S − T + I	3**C**	7**H**				18ζ
=3=	+ V − I + L	3**C**	7**H**				18ζ
=3=	+ N + I − Q	3**C**	7**H**				19ζ
=3=	+ S − T + L	3**C**	7**H**				20ζ
=3=	+ N + L − Q	3**C**	7**H**				21ζ
=3=	+ G − A + I	3**C**	7**H**				23ζ
=3=	− A + I − *	3**C**	6**H**		−1*		23ζ
=3=	+ G − A + L	3**C**	7**H**				25ζ
=3=	− A + L − *	3**C**	6**H**		−1*		25ζ
=3=	+ A − C + M	3**C**	7**H**				35ζ
=4=	− S + P + T − ϕ	4**C**	7**H**			−1ϕ	8ζ
=4=	+ P − V + I − ϕ	4**C**	7**H**			−1ϕ	10ζ
=4=	+ P − V + L − ϕ	4**C**	7**H**			−1ϕ	12ζ
=4=	+ P − D + E − ϕ	4**C**	7**H**			−1ϕ	15ζ
=4=	− G + I + * − ϕ	4**C**	8**H**		1*	−1ϕ	17ζ
=4=	+ S − Q + K + ϕ	2**C**	7**H**			1ϕ	18ζ
=4=	− G + L + * − ϕ	4**C**	8**H**		1*	−1ϕ	19ζ
=4=	+ A − P + I + ϕ	2**C**	7**H**			1ϕ	27ζ
=4=	+ A − P + L + ϕ	2**C**	7**H**			1ϕ	29ζ
Aspartic Acid D {3⁺ 8⁻}							
=3=	+ D + I − L	2**C**	3**H**	2**O**			15ζ
=3=	+ D − I + L	2**C**	3**H**	2**O**			19ζ
=3=	+ V − L + E	2**C**	3**H**	2**O**			20ζ
=3=	+ V − I + E	2**C**	3**H**	2**O**			22ζ
=3=	+ S − T + E	2**C**	3**H**	2**O**			24ζ
=3=	+ N − Q + E	2**C**	3**H**	2**O**			25ζ
=3=	+ G − A + E	2**C**	3**H**	2**O**			29ζ

Contd.

Table 4.4 Contd

=3=	− A + E − *	2C	2H	2O		−1*		29ζ
=4=	+ S − F + Y + ϕ	1C	3H	2O			1ϕ	15ζ
=4=	− G + E + * − ϕ	3C	4H	2O		1*	−1ϕ	23ζ
=4=	+ A − P + E + ϕ	1C	3H	2O			1ϕ	33ζ
Asparagine N {5⁺ 7⁻}								
=3=	+ D + Q − E	2C	4H	1O	1N			15ζ
=3=	+ V − L + Q	2C	4H	1O	1N			18ζ
=3=	+ V − I + Q	2C	4H	1O	1N			20ζ
=3=	+ N + I − L	2C	4H	1O	1N			21ζ
=3=	+ S − T + Q	2C	4H	1O	1N			22ζ
=3=	+ G − A + Q	2C	4H	1O	1N			27ζ
=3=	− A + Q − *	2C	3H	1O	1N	−1*		27ζ
=4=	+ P − V + Q − ϕ	3C	4H	1O	1N		−1ϕ	14ζ
=4=	+ S − L + K + ϕ	1C	4H	1O	1N		1ϕ	20ζ
=4=	− G + Q + * − ϕ	3C	5H	1O	1N	1*	−1ϕ	21ζ
=4=	+ S − I + K + ϕ	1C	4H	1O	1N		1ϕ	22ζ
=4=	+ A − P + Q + ϕ	1C	4H	1O	1N		1ϕ	31ζ
Isoleucine I {3⁺3⁻}								
=3=	+ V − N + Q	4C	9H					22ζ
=P=	− S + T + V	4C	9H					23ζ
=3=	+ V − D + E	4C	9H					30ζ
=4=	+ G + V − * + ϕ	3C	8H			−1*	1ϕ	24ζ
=4=	+ S − N + K + ϕ	3C	9H				1ϕ	24ζ
=I=	+ T − Q + K + ϕ	3C	9H				1ϕ	25ζ
Leucine L {2⁺ 4⁻}								
=T=	+ G + A + P	4C	9H					11ζ
=T=	+ A + P − *	4C	8H			−1*		11ζ
=3=	− G + A + V	4C	9H					18ζ
=3=	+ A + V + *	4C	10H			1*		18ζ
=4=	− A + P + V − ϕ	5C	9H				−1ϕ	14ζ
=L=	+ P − C + M − ϕ	5C	9H				−1ϕ	33ζ

Leucine and isoleucine configurations are determined in a mixed gauge 4^+2^- and 3^+3^- of amino acids conformal field $\phi(I, L)\ \{5^+7^-\} = \{\phi^*, \phi^{-*}\}$; (I, L) is a marked pair of field ϕ.

Let us introduce ζ-charge of the amino acid $Q_\zeta(Am)$, which differentiates dominant states of the amino acid in the field **GF**(83). Then difference of multiple charges

$$Q_\zeta\{I\}\begin{Bmatrix}5^+\\3^+\ 3^-\end{Bmatrix}-Q_\zeta\{L\}\begin{Bmatrix}7^-\\2^+\ 4^-\end{Bmatrix}=Q_p(H)\{0^+2^-\}$$

is in concordance with metric of the amino acid H dominant states. It proves correctness of distribution of configurations between amino acids I and L. The connection symbols, introduced here for the coincident charges $Q_p(\alpha)$ and $Q_p(\beta)$:

$$\underset{\Gamma\alpha\beta}{Q}=\begin{Bmatrix}Q_\phi\\\alpha^+\ \beta^-\end{Bmatrix}$$

of the amino acids α and β perform differentiation of amino acids by the character of field ϕ.

Let us rearrange configurations I, L and form the fields **GF**(83), **GF**(167) by using ζ-charge of amino acids. Then select the configurations of informational and basic DNA and RNA nucleotides in these fields. We give two CFAA variants from the number of possible ones.

$\phi\,(4^+2^-)$

			GF$_{\zeta*}$ (83)			
A70ζ						
=T=	$+G+A+P$	4C	9H			11ζ
=T=	$+A+P-*$	4C	8H	–1*		11ζ
=3=	$-G+A+V$	4C	9H			18ζ
=3=	$+V-D+E$	4C	9H			30ζ
G78ζ						
=4=	$-A+P+V-\phi$	5C	9H		-1ϕ	14ζ
			GF$_{\zeta**}$ (167)			
=3=	$+A+V+*$	4C	10H	1*		18ζ
=3=	$+V-N+Q$	4C	9H			22ζ
=4=	$+G+V-*+\phi$	3C	8H	–1*	1ϕ	24ζ
C47ζ						
=P=	$-S+T+V$	4C	9H			23ζ
=4=	$+S-N+K+\phi$	3C	9H		1ϕ	24ζ
U58ζ						
=I=	$+T-Q+K+\phi$	3C	9H		1ϕ	25ζ
=L=	$+P-C+M-\phi$	5C	9H		-1ϕ	33ζ

$\phi\,(3^+3^-)$

GF$_{\zeta*}$ (83)

A70ζ								
=3=	+ V − N + Q	4C	9H					22ζ
=4=	+ G + V − * + ϕ	3C	8H			−1*	1ϕ	24ζ
=4=	+ S − N + K + ϕ	3C	9H				1ϕ	24ζ
U58ζ								
=4=	− A + P + V − ϕ	5C	9H				−1ϕ	14ζ
			$GF_{\zeta **}$ (167)					
=**T**=	+ A + P − *	4C	8H			−1*		11ζ
=L=	+ P − C + M − ϕ	5C	9H				−1ϕ	33ζ
G78ζ								
=P=	− S + T + V	4C	9H					23ζ
=3=	+ V − D + E	4C	9H					30ζ
=**I**=	+ T − Q + K + ϕ	3C	9H				1ϕ	25ζ
C47ζ								
=**T**=	+ G + A + P	4C	9H					11ζ
=3=	− G + A + V	4C	9H					18ζ
=3=	+ A + V + *	4C	10H			1*		18ζ

Glutamic acid E $\{5^+ 7^-\}$

=3=	− G + A + D	3C	5H	2O				19ζ
=3=	+ A + D + *	3C	6H	2O		1*		19ζ
=3=	− S + T + D	3C	5H	2O				24ζ
=3=	− V + D + I	3C	5H	2O				26ζ
=3=	− V + D + L	3C	5H	2O				28ζ
=3=	+ I − L + E	3C	5H	2O				29ζ
=3=	− I + L + E	3C	5H	2O				33ζ
=4=	− A + P + D − ϕ	4C	5H	2O			−1ϕ	15ζ
=4=	− A + S + T + ϕ	2C	5H	2O			1ϕ	18ζ
=4=	+ T − F + Y + ϕ	2C	5H	2O			1ϕ	22ζ
=4=	+ G + D − * + ϕ	2C	4H	2O		−1*	1ϕ	25ζ
=4=	− P + V + D + ϕ	2C	5H	2O			1ϕ	32ζ

Glutamine Q $\{5^+ 7^-\}$

=3=	− G + A + N	3C	6H	1O	1N			25ζ
=3=	+ A + N + *	3C	7H	1O	1N	1*		25ζ
=3=	− S + T + N	3C	6H	1O	1N			30ζ
=3=	− I + L + Q	3C	6H	1O	1N			31ζ
=3=	− V + N + I	3C	6H	1O	1N			32ζ

=3=	− V + N + L	3C	6H	1O	1N			34ζ
=3=	− D + N + E	3C	6H	1O	1N			37ζ
=4=	− A + P + N − ϕ	4C	6H	1O	1N		−1ϕ	21ζ
=4=	+ T − I + K + ϕ	2C	6H	1O	1N		1ϕ	29ζ
=4=	+ G + N − * + ϕ	2C	5H	1O	1N	−1*	1ϕ	31ζ
=4=	+ S − V + K + ϕ	2C	6H	1O	1N		1ϕ	31ζ
=4=	− P + V + N + ϕ	2C	6H	1O	1N		1ϕ	38ζ

Lysine K {7⁺ 3⁻}

=3=	+ T − Y + W	4C	10H	1N		28ζ
=3=	+ I − L + K	4C	10H	1N		35ζ
=3=	− I + L + K	4C	10H	1N		39ζ
=4=	+ P − F + W − ϕ	5C	10H	1N	−1ϕ	23ζ
=4=	+ P − H + R + ϕ	3C	10H	1N	1ϕ	25ζ
=4=	− S + V + Q − ϕ	5C	10H	1N	−1ϕ	35ζ
=4=	− T + I + Q − ϕ	5C	10H	1N	−1ϕ	37ζ
=4=	− S + N + I − ϕ	5C	10H	1N	−1ϕ	38ζ
=4=	− T + L + Q − ϕ	5C	10H	1N	−1ϕ	39ζ
=4=	− S + N + L − ϕ	5C	10H	1N	−1ϕ	40ζ

Methionine M {4⁺ 3⁻}

=3=	+ I − L + M	3C	7H	1S			39ζ
=3=	− I + L + M	3C	7H	1S			43ζ
=3=	− A + C + V	3C	7H	1S			22ζ
=4=	+ G + A + C + ϕ	2C	7H	1S		1ϕ	19ζ
=4=	+ A + C − * + ϕ	2C	6H	1S	−1*	1ϕ	19ζ
=4=	− P + C + I + ϕ	2C	7H	1S		1ϕ	33ζ
=4=	− P + C + L + ϕ	2C	7H	1S		1ϕ	35ζ

Histidine H {0⁺ 2⁻}

=3=	+ I − L + H	4C	5H	2N	47ζ
=3=	− I + L + H	4C	5H	2N	51ζ

Phenylalanine F {4⁺ 3⁻}

=3=	+ I − L + F	7C	7H		57ζ
=3=	− I + L + F	7C	7H		61ζ
=3=	+ A − S + Y	7C	7H		63ζ
=4=	+ P − K + W − ϕ	8C	7H	−1ϕ	45ζ

=4=	$+T-E+Y+\phi$	6C	7H			1ϕ	50ζ
=4=	$+P-T+Y-\phi$	8C	7H			-1ϕ	54ζ
=4=	$+S-D+Y+\phi$	6C	7H			1ϕ	57ζ

Arginine R $\{1^+2^-\}$

=3=	$+I-L+R$	4C	10H	3N		59ζ
=3=	$-I+L+R$	4C	10H	3N		63ζ
=4=	$-P+K+H-\phi$	5C	10H	3N	-1ϕ	73ζ

Tyrosine Y $\{3^+4^-\}$

=3=	$+T-K+W$	7C	7H	1O		55ζ
=3=	$-A+S+F$	7C	7H	1O		60ζ
=3=	$+I-L+Y$	7C	7H	1O		62ζ
=3=	$-I+L+Y$	7C	7H	1O		66ζ
=4=	$-S+D+F-\phi$	8C	7H	1O	-1ϕ	66ζ
=4=	$-P+T+F+\phi$	6C	7H	1O	1ϕ	69ζ
=4=	$-T+E+F-\phi$	8C	7H	1O	-1ϕ	73ζ

Tryptophan W $\{1^+3^-\}$

=3=	$+I-L+W$	9C	12H	1N		79ζ
=3=	$-I+L+W$	9C	12H	1N		83ζ
=3=	$-T+K+Y$	9C	12H	1N		90ζ
=4=	$-P+K+F+\phi$	8C	12H	1N	1ϕ	95ζ

It follows from Table 4.4 that the interval (–1, +1) = (*, G) is divided asymmetrically when differentiating coincident charges, but CFAA metric remains invariant

$$*/G = (6^+4^-) / (10^+7^-) = (+6^+ - 7^-, +4^- - 10^+) = 13^+7^-.$$

In the field **GF**(83), it is necessary to calculate the exchanging charges of the corresponding configurations when building instanton site from paired (doubled) charges $Q_p(Am)$, $Q_\zeta(Am)$, as it has been performed for 10-parametric Lie group. Then it is necessary to identify p- and ζ-charges and to replace the large exchanging charges by superposition of ζ-charges, i.e., by the fragments of line-connected strands of amino acids. Each selected configuration of charge $Q_\zeta(Am)$ in the field **GF**(83) has its proper metric and, therefore, it is possible to build a great number of equivalent instanton configurations. No matter how the differentiation of charge configurations was performed, it is not possibly to determine unambiguously what charge configuration should be used as an exchange charge, because charge-exchanging group is infinite, isomorphic to the Braid group and contains symmetric group as a factor group. Charge-exchanging group is built on doubled charges and, therefore, principally contains a set of equivalent charge configurations.

Identification of p- and ζ-charges of the amino acid Am in the field **GF**(83) is similar to the identification of inert and gravitational body mass in the general theory of relativity. However, for each p-charge, ζ-charge is determined in each CFAA point and both charges are not equal to each other numerically. This property makes CFAA considerably different from the classic gravitational field. As a result of p- and ζ-charges pairing, the doubled Riemann space is formed and, therefore, CFAA possesses duplication properties. The bundle of terminator * and amino acid G forms correspondingly a 10-dimensional and 17-dimensional tangent spaces and 17-parametric and 10-parametric movement groups. Involution element which transfers the terminator* space into the space of amino acid G is CFAA-forming element (metric tensor) with metric $13^{+}7^{-}$ and 20-parametric movement group.

Representation of information in a conformal field

Charge-exchanging group which is necessary for the binding of primary instanton junctions, performs not only exchange of protonic charges of amino acids between different CFAA charge configurations but also transfers information between the states of field ϕ, which are expressed through the basic amino acids of the genetic code.

State CFAA ϕ^{+2} and ϕ^{-2} are copies of states ϕ^{+1} and ϕ^{-1}; CFAA states ϕ and $\phi5$ are correspondingly principal and dual basic state.

Let us determine CFAA state in the following way:

ϕ^{-1}:

$$\delta\phi \neq 0; \delta Q_{p\zeta} = Q_p(Am) - Q_\zeta(Am) = 0;$$

$$\delta\phi = 0; \delta Q_{p\zeta} \neq 0$$

ϕ^{+1}:

$$\delta\phi = 0; \delta Q_{p\zeta} = Q_p(Am) - Q_\zeta(Am) = 0;$$

$$\delta\phi \neq 0; \delta Q_{p\zeta} \neq 0$$

ϕ^{-2}: $$\delta\phi = 0; \delta Q_{p\zeta} = Q_p(Am) - Q_\zeta(Am) = 0;$$

ϕ^{+2}: $$\delta\phi \neq 0; \delta Q_{p\zeta} \neq 0$$

State of $\phi 5$ will be determined later.

Each of the states, ϕ^{+1} and ϕ^{-1}, can be expressed through DNA informational nucleotides. State ϕ^{-1} determines the unclosed DNA structure; state ϕ^{+1} corresponds to circular DNA (it is also possible to associate states ϕ^{-1} and ϕ^{+1} with DNA introns and exons).

State ϕ^{-1} of field ϕ is characterized by a negative inverse bond on exchange charge $\delta Q_{p\zeta}$ and is, therefore, related to the stable state, while the state ϕ^{+1} is unstable, transitive and auto-inductive.

States ϕ^{+2} and ϕ^{-2} are expressed through RNA informational nucleotides and can be associated with primary RNA transcript and informational RNA.

There is a link between the states. In order to find it, let us select the dominant states of amino acids in the field **GF**(83), for which an identity $\delta Q_{p\zeta} = 0$ is true. Let us obtain the CFAA exchanging set between states ϕ^{+1} and ϕ^{-1}:

$$S(17), P(23), P(23), P(23), V(25), V(25), N(31), L(33), \phi.$$

Constraint equation looks like:

$$Q_4\{+S(17) - P(23) - P(23) - P(23) + V(25) + V(25) - N(31) + L(33)\}_\phi = 0.$$

The set of exchanging amino acids is a radical of field ϕ, i.e., set of irreversible CFAA elements: coincident exchanging charges P(23) and V(25) are included in the constraint equation. Since $Q_4\{\phi\} = \mathbf{Z}_2$, then exchanging set of amino acids is the carrier of incident charges h(Am), which are bound to each other by the amino acid strand. Thus, exchanging set of amino acids is an informational message. It is clear from here that any 0-vector, which contains the field ϕ with condition $Q_4\{0\}_\phi = 0$ is also an informational exchanging vector between CFAA states. Let us transform an exchanging set of amino acids into the set of DNA informational nucleotides (Fig. 4.7).

Informational exchange vector ϕ can be transformed into a standard connection form as a factor space of CFAA curvature tensor with a *function* of informational nucleotide **C** attachment. In calibration $\phi = 0$, the connection looks like:

$$\Gamma^{A}_{CG} = \begin{Bmatrix} \mathbf{A} & Q_4\{0\}_\phi \\ \mathbf{C} & \mathbf{G} \end{Bmatrix}$$

In order to pass correctly from the connection to the CFAA curvature tensor, it is necessary to determine the informational nucleotide **T**, which has to be complementary to nucleotide **A**. The sense of complementarity lies

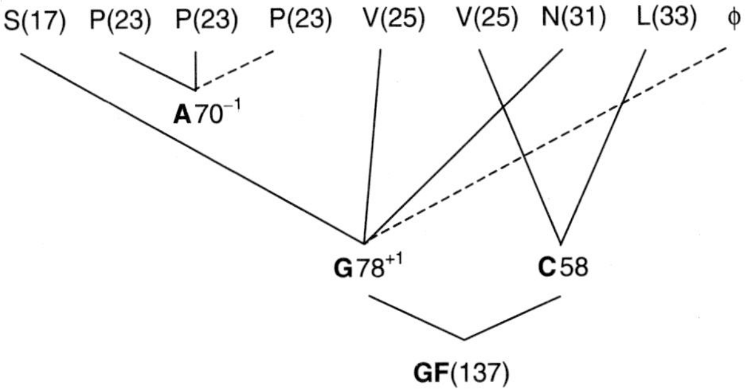

Fig. 4.7 Representation of charge-exchanging group by the set of DNA informational nucleotides. Complementary pair **GC** is formed as a result of transfer of the message_vector ϕ from **G** to **C**. The **GC** pair is a bundle of field **GF**(137) – inverse fine structure constant. Introduction of informational nucleotide **T** requires expansion of field **GF**(137) up to **GF**(167). Informational nucleotides are designated as a state of field ϕ.

in the exchange of metric tensor of amino acids conformal field between DNA and RNA; otherwise, the basic nucleotide **T** would have been completely equivalent to the basic nucleotide U when exchanging **T** $\leftrightarrow$ **U** (involutive CFAA metric tensor $10^{+}\,10^{-}$ during DNA $\leftrightarrow$ RNA transition is examined in Chapter 5).

Let us transform exchanging set of amino acids into the field **GF**(167) by introduction of mixed calibration of amino acid charge. One part of the amino acids will be determined by charge $Q_p(\mathrm{Am})$, the remaining part will be determined in the field **GF**(83) by charge $Q_\zeta(\mathrm{Am})$. Let us introduce the marked point (**I**, **L**) of field **GF**(83) instead of the marked point (**E**, **Q**). Transformed set of amino acids is a field **GF**(167):

$$Q_p(\mathrm{PPP}) \rightarrow Q_\zeta(\mathrm{CTV}); Q_p(\mathrm{VV}) \rightarrow Q_\zeta(\mathrm{EQ}); Q_\zeta(\mathbf{L}) \rightarrow Q_\zeta(\mathbf{I});$$
$$\mathrm{C}(9) + \mathrm{T}(11) + \mathrm{V}(16^*) + \mathrm{E}(29) + \mathrm{Q}(31) + \mathbf{T66} + \phi 5(\mathbf{S})^{+1} = \mathbf{GF}(167);$$

where $\mathbf{T66}_{\phi 5} = \mathbf{I}(25) + \mathrm{N}(31)$ and chemical element **S** are the marked points of fields $\phi 5$ and $\phi = \mathbf{GF}(167)$ correspondingly.

Curvature of field ϕ is determined by charge $Q_p(\mathbf{T})$:

$$Q_p(\mathbf{T66^*}) \rightarrow Q_p\{\mathrm{T}(25)\}; Q_p(\mathrm{C, T, V, E, Q, T})_{\phi 5} = \mathbf{GF}(167)$$

with promoter basis $\mathrm{M}(41)^* = Q_p\{\mathbf{T66^*} - \mathrm{T}(25)\}$.

We also have

$$Q_4\{+\mathrm{M}(41)^* + [\mathrm{C}_\zeta(9) \rightarrow + \mathbf{A}_\zeta(3)] - \mathrm{T}(11) +$$
$$\mathrm{V}(16^*) - \mathrm{E}(29) - \mathrm{Q}(31) + \mathbf{I}(25) - \mathrm{N}(31) + \mathrm{S}(17)\}_\phi = 0,$$

where $\mathbf{A}_\zeta(3)$ is the marked point of field **GF**(167). The obtained formula is also an informational message with the controlling operator of transition $C_\zeta(9)_\phi \rightarrow \mathbf{A}_\zeta(3)$, which marks amino acid A by changing the evenness of its state.

Informational nucleotide **T66*** is formed by the bundle of basic nucleotide $T55_{-2}$ via twisting of bundle base: the dual basis $\phi 5$ is turned into ϕ by twisting $T(25) \rightarrow C(25)$. Curvature tensor of the field ϕ looks like:

$$R_{CG}^{A\phi} = \begin{vmatrix} \mathbf{A} & M(41) * \mathbf{T66}_{C(25)} \\ \mathbf{C} & \mathbf{G} \end{vmatrix},$$

where identity of complementarity is fulfilled:

$$[\mathbf{CG}] = [\mathbf{AT}] = M(41) + \mathbf{T66} + V(25) + \phi 5 = \mathbf{GF}(137),$$

and transition $C(25) \rightarrow V(25)$ turns the basis ϕ into the basis $\phi 5$. Informational nucleotide **T** is the basis of $\phi 5$.

The duration of CFAA stay in an unstable state ϕ^{+1}, which is marked by the accepted vector of informational message, is limited by metric tensor of the informational message vector itself, the negative part of which determines the exchanging charge with another CFAA point, to which the state ϕ^{+1} passes.

Representation of entropy Ω^+

Let us arrange the state of the genetic code from the left to the right by protonic charge of amino acids starting with the charge of terminator * equal to –1. Let us write in a row the states of each amino acid; amino acids with coincident charges will be placed one after another in an alphabetic order. Let us obtain the primary arrangement of amino acids (Table 4.5) of the ordinal type ω_1. Then, starting from charge –1 and further as far as charge increases from the right to the left, let us rearrange the coincident charge accordingly to the following principle: first charge from the right will be invariable, every next charge to the left will be considered as a copy of preceding one and receives increment +2, +3… . Let us obtain the resequencing of amino acids, which belong to the ordinal type ω_2.

All charges of amino acids are different now. Let us arrange the new charges in an ascending order with preservation of ω_2. We obtain the arrangement of amino acids by type ω_3, which is the representation of entropy Ω^+ of the genetic code with introns. The obtained charges of the genetic code amino acids can be accepted as the dominant charges in the field **GF**(83).

In order to obtain the equality of ordinal numbers **Q**2 = **Q**3, which expresses the identity of marked points by equivalent states of genetic code, it is necessary to get rid of the introns. This operation is known as RNA splicing.

Isolated charges form field **GF**(167):

$$Q_p\{A(9) + S(17) + S(71) + W(73)\}^{***} = \mathbf{GF}(167).$$

Let us give the charge –1 to introns of the ordinal type ω_3 (1, 2, 1, 3, 1, 1, 1, 1) and rearrange the charges of amino acids:

A*AAAG_GSGSSSSPGVPVVVTNTLDLLLLQIMEHKHKQL
FIFEINRTYRYRRRRDTCPCP_S_W** = {+9, +10, ..., +73**}.

The field of isolated charges decreases to

$$Q_p\{S(71) + W(73)^{**}\} = 2\mathbf{GF}(71).$$

Table 4.5 Representation of entropy $\Omega^+ = 74$ with introns

*	*	*	G	G	G	G	A	A	A	A	S	S	S	S	S
S	P	P	P	P	C	C	T	T	T	T	V	V	V	V	D
D	N	N	I	I	I	L	L	L	L	L	L	E	E	Q	Q
K	K	M	H	H	F	F	R	R	R	R	R	R	Y	Y	W
−1	−1	−1	1	1	1	1	9	9	9	9	17	17	17	17	17
17	23	23	23	23	25	25	25	25	25	25	25	25	25	25	31
31	31	31	33	33	33	33	33	33	33	33	33	39	39	39	39
41	41	41	43	43	49	49	55	55	55	55	55	55	57	57	73
2	1	−1	1	1	1	1	9	9	9	9	17	17	17	17	17
17	23	23	23	23	25	25	25	25	25	25	25	25	25	25	31
31	31	31	33	33	33	33	33	33	33	33	33	39	39	39	39
41	41	41	43	43	49	49	55	55	55	55	55	55	57	57	73
2	6	−1	5	4	3	1	9	9	9	9	17	17	17	17	17
17	23	23	23	23	25	25	25	25	25	25	25	25	25	25	31
31	31	31	33	33	33	33	33	33	33	33	33	39	39	39	39
41	41	41	43	43	49	49	55	55	55	55	55	55	57	57	73
2	6	−1	5	4	3	1	13	12	11	9	17	17	17	17	17
17	23	23	23	23	25	25	25	25	25	25	25	25	25	25	31

Contd.

Table 4.5 Contd

31	31	31	33	33	33	33	33	33	33	33	33	39	39	39	39
41	41	41	43	43	49	49	55	55	55	55	55	55	57	57	73
.	.	.	.	.	.	.	.	.	.	.	.	.	.	.	.
2	6	−1	5	4	3	1	13	12	11	9	69	22	21	20	19
17	69	26	67	23	68	66	65	32	56	30	29	28	27	25	64
34	54	31	50	53	40	48	38	37	36	35	33	52	42	47	39
44	46	41	45	43	51	49	63	62	61	58	60	55	59	57	73
2	6	−1	5	4	3	1	13	12	11	9	71	22	21	20	19
17	69	26	67	23	68	66	65	32	56	30	29	28	27	25	64
34	54	31	50	53	40	48	38	37	36	35	33	52	42	47	39
44	46	41	45	43	51	49	63	62	61	58	60	55	59	57	73

*_G*GGG*__A_AAA___S_SSSSP_VPVVVTNTLDLLLLQIMEHKHKQL

FIFEINRTYRYRRRRDTCPCP_S_W = {−1, +1, +2, ... ,73}

Introns are labelled as underlined symbols.

The field of introns is $Q_p\{16^* + 70 + 72\} = 16^* + 2\mathbf{GF}(71)$. The binding intron G_G provides an entropy production and is therefore, cut off; the field **GF**(71) turns into connection and the process of amino acid sequence compression stops. The new rearrangement of the genetic code amino acids

A*AAAGGSGSSSSPGVPVVVTNTLDLLLLQIMEHKHKQL
FIFEINRTYRYRRRRDTCPCP*S*W = {+10*, +11, ..., +73}.

is simple and superentropic.

Equivalent Charge Configurations

5
Chapter

Dominant configurations of nucleotides in DNA and RNA

In Chapter 4, we examined 3- and 4_ϕ-charge configurations of the genetic code amino acids. The existence of the ideal ±(I–L) is necessary and a sufficient condition for representation of amino acid configurations by indexes 3…19 and $4_\phi \ldots 20_\phi$. These configurations are the coherent bundle of the genetic code amino acids and, consequently, determine the covariant derivative of protein, where each amino acid is identical to the potential of a conformal field. However, the basic nucleotides as well as the informational ones of DNA and RNA are also the charge configurations of amino acids.

All charge configurations of informational nucleotides except thymine begin with 5- and 6_ϕ-index configurations. The minimum configurations of adenine and guanine are the 6_ϕ-index configurations that are complementary to minimum three-, five-index thymine configurations and to five-index cytosine configurations.

The fact that basic and informational DNA and RNA nucleotides are expressed via linear combinations of the genetic code amino acids, it proves that there are two connection forms of the amino acids conformal field. We can consider the protein that contains more than three different amino acids as a dynamic memory of all amino acids configurations, and all nucleotide triplets of the genetic code. It explains the ability of protein to perform various biological functions.

All 3- and 4_ϕ-index configurations of amino acids form the true protein connection, whereas configurations of higher order are the copies of amino acids. DNA and RNA are, as it is known, the particular case of protein copying (preservation), but in a six-dimensional space. Therefore, protein does not store the self-copy. All amino acid configurations, which obviously contain the basis of field ϕ, take into account the curvature of CFAA Riemann space. DNA and RNA informational nucleotides, which contain seven indexes, form the flat tangent space in each CFAA point and, therefore, the connection of informational nucleotides determines one amino acid of genetic code modulo 20-parametric Lie group. The genetic code can be considered as an algorithm of CFAA connection reduction to zero for each protein amino acid, i.e., the genetic code degeneracy is necessary for irreversible transformation of RNA into protein. The connection of nine-index configurations of DNA and RNA informational nucleotides is again reduced to seven-index configurations of amino acids modulo 20-parametric Lie group, i.e., to the copies of informational nucleotides; therefore, DNA and RNA possess a property of self-reproduction. CFAA metric $13^+\ 7^-$ determines a process of semi-conservative DNA self-reproduction on the connection basis of 13-index configurations of nucleotides but modulo 10-parametric Lie group with consecutive self-duplication of nine-index configurations connection. Thus, nine-index nucleotide configurations are the potential of DNA and RNA self-reproduction.

19-Index nucleotide configurations also possess a property of self-reproduction. As $13^+ = 19_C{}^+$, thus, in this case field ϕ absorbs its own value $+1\mathbf{C}$ and forms its copies: $+1\mathbf{C} +2\mathbf{H} = \phi^{+2}$, $-1\mathbf{C} +2\mathbf{H} = \phi^{-2}$ additionally to the basic state $\pm1\mathbf{C} = \phi^{\pm1}$. Local field $\phi^{-1}, \phi^{-2}, \phi^{+1}, \phi^{+2}$ has a spin metric $2^+\,2^-$ with the basis $\mathbf{CH_4} \rightarrow \mathbf{CO}$ of parallel transfer of the marked point $\mathbf{OH_4}$ of field $\phi\{4^+\,2^-\}$. The marked point $\mathbf{OH_4}$ is formed in DNA via connecting of the informational nucleotide to the mark 2' of a furanose ring.

Absorption of amino acids with its own basis +1C by a conformal field in the absence of any other modes of field expansion, leads to one of the possible ways of cell malignization.

Table 5.1 represents charge configurations of basic, informational and some other minor DNA and RNA nucleotides.

Table 5.1

		Basic Uracil $U45_{-1*}$						
=3=	–S+P+E	5**C**	7**H**	1**O**				34ζ
=4=	–G+S+P+φ	4**C**	7**H**	1**O**			1φ	16ζ
=4=	+S+P+*+φ	4**C**	8**H**	1**O**		1*	1φ	16ζ
=4=	–A+P+T+φ	4**C**	7**H**	1**O**			1φ	21ζ
=4=	–S+V+D+φ	4**C**	7**H**	1**O**			1φ	35ζ
=4=	–T+D+I+φ	4**C**	7**H**	1**O**			1φ	37ζ
=4=	–T+D+L+φ	4**C**	7**H**	1**O**			1φ	39ζ
=4=	–T+V+E+φ	4**C**	7**H**	1**O**			1φ	42ζ
=4=	–P+T+F–φ	6**C**	7**H**	1**O**			–1φ	57ζ
		Basic Thymine $T55_{-2}$						
=3=	–G+P+D	5**C**	7**H**	2**O**				23ζ
=3=	+P+D+*	5**C**	8**H**	2**O**		1*		23ζ
=3=	–A+P+E	5**C**	7**H**	2**O**				35ζ
=4=	–A+V+D+φ	4**C**	7**H**	2**O**			1φ	36ζ
=4=	–G+A+E+φ	4**C**	7**H**	2**O**			1φ	39ζ
=4=	+A+E+*+φ	4**C**	8**H**	2**O**		1*	1φ	39ζ
=4=	–N+Q+E+φ	4**C**	7**H**	2**O**			1φ	43ζ
=4=	–S+T+E+φ	4**C**	7**H**	2**O**			1φ	44ζ
=4=	–V+I+E+φ	4**C**	7**H**	2**O**			1φ	46ζ
=4=	–V+L+E+φ	4**C**	7**H**	2**O**			1φ	48ζ
=4=	–C+D+M+φ	4**C**	7**H**	2**O**			1φ	55ζ
=4=	–P+T+Y–φ	6**C**	7H	2**O**			– 1φ	62ζ
		Conjugated Basic Cytosine C47						
=5=	–A+D+H+R–W		3**H**	2**O**	4**N**			43ζ
=5=	+E–K+H–F+R		3**H**	2**O**	4**N**			45ζ
=5=	+S–P+N–L+R		3**H**	2**O**	4**N**			54ζ
=5=	+S–P+N–I+R		3**H**	2**O**	4**N**			56ζ
=6=	+T–I–L+Q+R–φ	1**C**	3**H**	2**O**	4**N**		–1φ	43ζ
=6=	+S–V–L+Q+R–φ	1**C**	3**H**	2**O**	4**N**		–1φ	45ζ
=6=	+T–V+N–L+R–φ	1**C**	3**H**	2**O**	4**N**		–1φ	46ζ
=6=	+S–V–I+Q+R–φ	1**C**	3**H**	2**O**	4**N**		–1φ	47ζ
=6=	+T–V+N–I+R–φ	1**C**	3**H**	2**O**	4**N**		–1φ	48ζ
		Conjugated Basic Uracil $U46_{\phi}$						
=4=	–P+D–L+R	–1**C**	–1**H**	2**O**	3**N**			44ζ
=4=	–P+D–I+R	–1**C**	–1**H**	2**O**	3**N**			46ζ
=5=	–V–L+E+R–φ		–1**H**	2**O**	3**N**		–1φ	43ζ
=5=	–V–I+E+R–φ		–1**H**	2**O**	3**N**		–1φ	45ζ

Contd.

Table 5.1 Contd

		Thymine T66*						
=3=	–A+D+H	5**C**	5**H**	2**O**	2**N**			63ζ
=4=	–P+E+H+ϕ	4**C**	5**H**	2**O**	2**N**		1ϕ	79ζ
=5=	+G+P+D–L+H	5**C**	5**H**	2**O**	2**N**			47ζ
=5=	–G+C+E–M+H	5**C**	5**H**	2**O**	2**N**			47ζ
=5=	+P+D–L+H–*	5**C**	4**H**	2**O**	2**N**	–1*		47ζ
=5=	+C+E–M+H+*	5**C**	6**H**	2**O**	2**N**	1*		47ζ
=5=	–G+T+Q–K+H	5**C**	5**H**	2**O**	2**N**			51ζ
=5=	+T+Q–K+H+*	5**C**	6**H**	2**O**	2**N**	1*		51ζ
=5=	–G+V+D–L+H	5**C**	5**H**	2**O**	2**N**			54ζ
=5=	+V+D–L+H+*	5**C**	6**H**	2**O**	2**N**	1*		54ζ
=5=	–G+V+D–I+H	5**C**	5**H**	2**O**	2**N**			56ζ
=5=	+V+D–I+H+*	5**C**	6**H**	2**O**	2**N**	1*		56ζ
=5=	–G+S–T+D+H	5**C**	5**H**	2**O**	2**N**			58ζ
=5=	+S–T+D+H+*	5**C**	6**H**	2**O**	2**N**	1*		58ζ
=5=	–G+D+N–Q+H	5**C**	5**H**	2**O**	2**N**			59ζ
=5=	+D+N–Q+H+*	5**C**	6**H**	2**O**	2**N**	1*		59ζ
=5=	–G+Q+H+Y–W	5**C**	5**H**	2**O**	2**N**			60ζ
=5=	+Q+H+Y–W+*	5**C**	6**H**	2**O**	2**N**	1*		60ζ
=5=	–A+D+I–L+H	5**C**	5**H**	2**O**	2**N**			61ζ
=5=	+G–A+D+H+*	5**C**	6**H**	2**O**	2**N**	1*		63ζ
=5=	–G–A+D+H–*	5**C**	4**H**	2**O**	2**N**	–1*		63ζ
=5=	–A+D–I+L+H	5**C**	5**H**	2**O**	2**N**			65ζ
=5=	+E–K+H–F+W	5**C**	5**H**	2**O**	2**N**			65ζ
=5=	–G+A–V+E+H	5**C**	5**H**	2**O**	2**N**			66ζ
=5=	–A+V–L+E+H	5**C**	5**H**	2**O**	2**N**			66ζ
=5=	+A–V+E+H+*	5**C**	6**H**	2**O**	2**N**	1*		66ζ
=5=	–S+D+H–F+Y	5**C**	5**H**	2**O**	2**N**			67ζ
=5=	–A+S–T+E+H	5**C**	5**H**	2**O**	2**N**			70ζ
=5=	–V–N+Q+E+H	5**C**	5**H**	2**O**	2**N**			70ζ
=5=	–A+N–Q+E+H	5**C**	5**H**	2**O**	2**N**			71ζ
=5=	–S+T–V+E+H	5**C**	5**H**	2**O**	2**N**			71ζ
=5=	+S–P+N–L+W	5**C**	5**H**	2**O**	2**N**			74ζ
=5=	–T+E+H–F+Y	5**C**	5**H**	2**O**	2**N**			74ζ
=5=	–A+V–I+E+H	5**C**	5**H**	2**O**	2**N**			68ζ
=5=	+S–P+N–I+W	5**C**	5**H**	2**O**	2**N**			76ζ
=5=	–C–V+D+M+H	5**C**	5**H**	2**O**	2**N**			82ζ
=5=	+S–P–K+R+Y	5**C**	5**H**	2**O**	2**N**			85ζ
=5=	–C–L+E+M+H	5**C**	5**H**	2**O**	2**N**			85ζ
=5=	–C–I+E+M+H	5**C**	5**H**	2**O**	2**N**			87ζ
=6=	–G+P+N–L+Q+ϕ	4**C**	5**H**	2**O**	2**N**		1ϕ	37ζ

Contd.

Table 5.1 Contd

=6=	+P+N–L+Q+*+ϕ	4**C**	6**H**	2**O**	2**N**	1*	1ϕ 37ζ
=6=	–G+P+N–I+Q+ϕ	4**C**	5**H**	2**O**	2**N**		1ϕ 39ζ
=6=	+P+N–I+Q+*+ϕ	4**C**	6**H**	2**O**	2**N**	1*	1ϕ 39ζ
=6=	–A+P–V+N+Q+ϕ	4**C**	5**H**	2**O**	2**N**		1ϕ 46ζ
=6=	+A+S+N–K+H+ϕ	4**C**	5**H**	2**O**	2**N**		1ϕ 48ζ
=6=	–G+P–V+D+H–ϕ	6**C**	5**H**	2**O**	2**N**		–1ϕ 50ζ
=6=	+P–V+D+H+*–ϕ	6**C**	6**H**	2**O**	2**N**	1*	–1ϕ 50ζ
=6=	–A+P+D–K+R+ϕ	4**C**	5**H**	2**O**	2**N**		1ϕ 51ζ
=6=	+G+S+Q–K+H+ϕ	4**C**	5**H**	2**O**	2**N**		1ϕ 52ζ
=6=	+S+Q–K+H–*+ϕ	4**C**	4**H**	2**O**	2**N**	–1*	1ϕ 52ζ
=6=	–G–A+N+Q+*+ϕ	4**C**	6**H**	2**O**	2**N**	1*	1ϕ 53ζ
=6=	–G+S+T–V+H+ϕ	4**C**	5**H**	2**O**	2**N**		1ϕ 53ζ
=6=	–G+P–L+E+H–ϕ	6**C**	5**H**	2**O**	2**N**		–1ϕ 53ζ
=6=	+G+T+N–K+H+ϕ	4**C**	5**H**	2**O**	2**N**		1ϕ 53ζ
=6=	+S+T–V+H+*+ϕ	4**C**	6**H**	2**O**	2**N**	1*	1ϕ 53ζ
=6=	+P–L+E+H+*–ϕ	6**C**	6**H**	2**O**	2**N**	1*	–1ϕ 53ζ
=6=	+T+N–K+H–*+ϕ	4**C**	4**H**	2**O**	2**N**	–1*	1ϕ 53ζ
=6=	–G+P–I+E+H–ϕ	6**C**	5**H**	2**O**	2**N**		–1ϕ 55ζ
=6=	+P–I+E+H+*–ϕ	6**C**	6**H**	2**O**	2**N**	1*	–1ϕ 55ζ
=6=	+G+A–V+D+H+ϕ	4**C**	5**H**	2**O**	2**N**		1ϕ 60ζ
=6=	+A–V+D+H–*+ϕ	4**C**	4**H**	2**O**	2**N**	–1*	1ϕ 60ζ
=6=	–S–T+D+N+Q+ϕ	4**C**	5**H**	2**O**	2**N**		1ϕ 60ζ
=6=	–A+S+H–F+Y+ϕ	4**C**	5**H**	2**O**	2**N**		1ϕ 61ζ
=6=	+G+N+H+Y–W+ϕ	4**C**	5**H**	2**O**	2**N**		1ϕ 62ζ
=6=	–A+P–V+E+H–ϕ	6**C**	5**H**	2**O**	2**N**		–1ϕ 62ζ
=6=	+N+H+Y–W–*+ϕ	4**C**	4**H**	2**O**	2**N**	–1*	1ϕ 62ζ
=6=	+G+A–L+E+H+ϕ	4**C**	5**H**	2**O**	2**N**		1ϕ 63ζ
=6=	+A–L+E+H–*+ϕ	4**C**	4**H**	2**O**	2**N**	–1*	1ϕ 63ζ
=6=	+T–I–L+Q+W–ϕ	6**C**	5**H**	2**O**	2**N**		–1ϕ 63ζ
=6=	+G+A–I+E+H+ϕ	4**C**	5**H**	2**O**	2**N**		1ϕ 65ζ
=6=	+A–I+E+H–*+ϕ	4**C**	4**H**	2**O**	2**N**	–1*	1ϕ 65ζ
=6=	+S–V–L+Q+W–ϕ	6**C**	5**H**	2**O**	2**N**		–1ϕ 65ζ
=6=	+T–V+N–L+W–ϕ	6**C**	5**H**	2**O**	2**N**		–1ϕ 66ζ
=6=	–G+A–P+D+H+ϕ	4**C**	5**H**	2**O**	2**N**		1ϕ 67ζ
=6=	+A–P+D+H+*+ϕ	4**C**	6**H**	2**O**	2**N**	1*	1ϕ 67ζ
=6=	+S–V–I+Q+W–ϕ	6**C**	5**H**	2**O**	2**N**		–1ϕ 67ζ
=6=	+T–V+N–I+W–ϕ	6**C**	5**H**	2**O**	2**N**		–1ϕ 68ζ
=6=	–G–A+E+H+*–ϕ	6**C**	6**H**	2**O**	2**N**	1*	–1ϕ 69ζ
=6=	–P+D–N+Q+H+ϕ	4**C**	5**H**	2**O**	2**N**		1ϕ 71ζ
=6=	+G–V+E+H–*+ϕ	4**C**	4**H**	2**O**	2**N**	–1*	1ϕ 72ζ
=6=	–S–P+T+D+H+ϕ	4**C**	5**H**	2**O**	2**N**		1ϕ 72ζ

Contd.

Table 5.1 Contd

=6=	−I−L+Q+K+Y−ϕ	6**C**	5**H**	2**O**	2**N**		−1ϕ	72ζ
=6=	−P−V+D+I+H+ϕ	4**C**	5**H**	2**O**	2**N**		1ϕ	74ζ
=6=	−V+N−L+K+Y−ϕ	6**C**	5**H**	2**O**	2**N**		−1ϕ	75ζ
=6=	−S−T+D+E+H−ϕ	6**C**	5**H**	2**O**	2**N**		−1ϕ	76ζ
=6=	−P−V+D+L+H+ϕ	4**C**	5**H**	2**O**	2**N**		1ϕ	76ζ
=6=	−P+I−L+E+H+ϕ	4**C**	5**H**	2**O**	2**N**		1ϕ	77ζ
=6=	+T−V−K+R+Y−ϕ	6**C**	5**H**	2**O**	2**N**		−1ϕ	77ζ
=6=	−V+N−I+K+Y−ϕ	6**C**	5**H**	2**O**	2**N**		−1ϕ	77ζ
=6=	+G−P+E+H+*+ϕ	4**C**	6**H**	2**O**	2**N**	1*	1ϕ	79ζ
=6=	−G−P+E+H−*+ϕ	4**C**	4**H**	2**O**	2**N**	−1*	1ϕ	79ζ
=6=	+D−L−Q+R+Y−ϕ	6**C**	5**H**	2**O**	2**N**		−1ϕ	80ζ
=6=	−P−I+L+E+H+ϕ	4**C**	5**H**	2**O**	2**N**		1ϕ	81ζ
=6=	−P−V+N+Q+F−ϕ	6**C**	5**H**	2**O**	2**N**		−1ϕ	82ζ
=6=	+D−I−Q+R+Y−ϕ	6**C**	5**H**	2**O**	2**N**		−1ϕ	82ζ
=6=	−P+D−K+F+R−ϕ	6**C**	5**H**	2**O**	2**N**		−1ϕ	87ζ

Adenine A70*

=6=	+P−V−L+H+R+ϕ	4**C**	4**H**		5**N**		1ϕ	80ζ
=6=	+P−V−I+H+R+ϕ	4**C**	4**H**		5**N**		1ϕ	82ζ
=6=	−G−L+H+R+*+ϕ	4**C**	5**H**		5**N**	1*	1ϕ	87ζ
=6=	−G−I+H+R+*+ϕ	4**C**	5**H**		5**N**	1*	1ϕ	89ζ
=6=	−T+N−K+H+R+ϕ	4**C**	4**H**		5**N**		1ϕ	91ζ
=6=	−G−A−V+H+R+ϕ	4**C**	4**H**		5**N**		1ϕ	96ζ
=6=	−A−V+H+R+*+ϕ	4**C**	5**H**		5**N**	1*	1ϕ	96ζ
=7=	−G−A−S+Q−K+H+R	5**C**	4**H**		5**N**			94ζ
=7=	−G−S−T−V+E+H+R	5**C**	4**H**		5**N**			109ζ
=7=	−G−S+N−K+H+R+*	5**C**	5**H**		5**N**	1*		90ζ
=7=	+G−T+N−I−L+R+W	5**C**	4**H**		5**N**			103ζ
=7=	−G−T+Q−K+H+R+*	5**C**	5**H**		5**N**	1*		89ζ
=7=	−A−S+Q−K+H+R+*	5**C**	5**H**		5**N**	1*		94ζ
=7=	−A−P−T+N−I+R+W	5**C**	4**H**		5**N**			119ζ
=7=	−A−P−T+N−L+R+W	5**C**	4**H**		5**N**			117ζ
=7=	−A−V−I−L+K+R+W	5**C**	4**H**		5**N**			108ζ
=7=	−S+P−V+N−K+H+R	5**C**	4**H**		5**N**			83ζ
=7=	−S−P−V−I+Q+R+W	5**C**	4**H**		5**N**			119ζ
=7=	−S−P−V−L+Q+R+W	5**C**	4**H**		5**N**			117ζ
=7=	−S+P−I+Q−K+H+R	5**C**	4**H**		5**N**			80ζ
=7=	−S+P−L+Q−K+H+R	5**C**	4**H**		5**N**			78ζ
=7=	−S−T−V+E+H+R+*	5**C**	5**H**		5**N**	1*		109ζ
=7=	+P−T−V+Q−K+H+R	5**C**	4**H**		5**N**			82ζ
=7=	−T+N−I−L+R+W−*	5**C**	3**H**		5**N**	−1*		103ζ

Contd.

Table 5.1 Contd

=7=	+N–I–L+Q–E+F+R	5C	4H		5N			89ζ
		Uracil U58*						
=5=	–G+C+D–M+H	4C	3H	2O	2N			33ζ
=5=	+C+D–M+H+*	4C	4H	2O	2N	1*		33ζ
=5=	–G+S+Q–K+H	4C	3H	2O	2N			44ζ
=5=	+S+Q–K+H+*	4C	4H	2O	2N	1*		44ζ
=5=	–G+T+N–K+H	4C	3H	2O	2N			45ζ
=5=	–A+C+E–M+H	4C	3H	2O	2N			45ζ
=5=	+T+N–K+H+*	4C	4H	2O	2N	1*		45ζ
=5=	+D–K+H–F+W	4C	3H	2O	2N			51ζ
=5=	–G+A–V+D+H	4C	3H	2O	2N			52ζ
=5=	–A+V+D–L+H	4C	3H	2O	2N			52ζ
=5=	+A–V+D+H+*	4C	4H	2O	2N	1*		52ζ
=6=	–A+P+N–L+Q+ϕ	3C	3H	2O	2N		1ϕ	35ζ
=6=	–A+P+N–I+Q+ϕ	3C	3H	2O	2N		1ϕ	37ζ
=6=	–G+P+D–L+H–ϕ	5C	3H	2O	2N		–1ϕ	39ζ
=6=	+P+D–L+H+*–ϕ	5C	4H	2O	2N	1*	–1ϕ	39ζ
=6=	–G+P+D–I+H–ϕ	5C	3H	2O	2N		–1ϕ	41ζ
=6=	+P+D–I+H+*–ϕ	5C	4H	2O	2N	1*	–1ϕ	41ζ
=6=	–G+S+T–L+H+ϕ	3C	3H	2O	2N		1ϕ	42ζ
=6=	+S+T–L+H+*+ϕ	3C	4H	2O	2N	1*	1ϕ	42ζ
=6=	–G+S+T–I+H+ϕ	3C	3H	2O	2N		1ϕ	44ζ
=6=	+S+T–I+H+*+ϕ	3C	4H	2O	2N	1*	1ϕ	44ζ
=6=	+G+S+N–K+H+ϕ	3C	3H	2O	2N		1ϕ	46ζ
=6=	+S+N–K+H–*+ϕ	3C	2H	2O	2N	–1*	1ϕ	46ζ
=6=	–A+P–V+D+H–ϕ	5C	3H	2O	2N		–1ϕ	48ζ
=6=	+G+A+D–I+H+ϕ	3C	3H	2O	2N		1ϕ	51ζ
=6=	–A+S+T–V+H+ϕ	3C	3H	2O	2N		1ϕ	51ζ
=6=	–A+P–L+E+H–ϕ	5C	3H	2O	2N		–1ϕ	51ζ
=6=	+A+D–I+H–*+ϕ	3C	2H	2O	2N	–1*	1ϕ	51ζ
=6=	+P–V+E–K+R+ϕ	3C	3H	2O	2N		1ϕ	52ζ
=6=	–A+P–I+E+H–ϕ	5C	3H	2O	2N		–1ϕ	53ζ
		Cytosine C58*						
=5=	–G+C+N–M+H	4C	4H	1O	3N			39ζ
=5=	+C+N–M+H+*	4C	5H	1O	3N	1*		39ζ
=5=	–A+C+Q–M+H	4C	4H	1O	3N			43ζ
=5=	–G+A–L+Q+H	4C	4H	1O	3N			53ζ
=5=	+A–L+Q+H+*	4C	5H	1O	3N	1*		53ζ
=5=	–G+A–I+Q+H	4C	4H	1O	3N			55ζ

Contd.

Table 5.1 Contd

=5=	–A+D+N–E+H	4C	4H	1O	3N		55ζ
=5=	+A–I+Q+H+*	4C	5H	1O	3N	1*	55ζ
=5=	+D–L+H–Y+W	4C	4H	1O	3N		56ζ
=5=	+N–K+H–F+W	4C	4H	1O	3N		57ζ
=5=	–G+A–V+N+H	4C	4H	1O	3N		58ζ
=5=	–A+V+N–L+H	4C	4H	1O	3N		58ζ
=5=	+A–V+N+H+*	4C	5H	1O	3N	1*	58ζ
=5=	–S+T–L+Q+H	4C	4H	1O	3N		58ζ
=5=	+D–I+H–Y+W	4C	4H	1O	3N		58ζ
=5=	–A+V+N–I+H	4C	4H	1O	3N		60ζ
=5=	–S+T–I+Q+H	4C	4H	1O	3N		60ζ
=5=	–V+I–L+Q+H	4C	4H	1O	3N		60ζ
=5=	+G–V+Q+H+*	4C	5H	1O	3N	1*	62ζ
=5=	–G–V+Q+H–*	4C	3H	1O	3N	–1*	62ζ
=5=	–A+S–T+N+H	4C	4H	1O	3N		62ζ
=5=	–S+T–V+N+H	4C	4H	1O	3N		63ζ
=5=	–V–I+L+Q+H	4C	4H	1O	3N		64ζ
=5=	–T+D–L+K+H	4C	4H	1O	3N		65ζ
=5=	–D–L+Q+E+H	4C	4H	1O	3N		65ζ
=5=	–T+N+H–F+Y	4C	4H	1O	3N		66ζ
=5=	–T+D–I+K+H	4C	4H	1O	3N		67ζ
=5=	+D–N–K+H+R	4C	4H	1O	3N		67ζ
=5=	–D–I+Q+E+H	4C	4H	1O	3N		67ζ
=5=	–G–P+Q+H+*	4C	5H	1O	3N	1*	69ζ
=5=	–V–D+N+E+H	4C	4H	1O	3N		70ζ
=5=	–Q+E–K+H+R	4C	4H	1O	3N		75ζ
=5=	–C+N–L+M+H	4C	4H	1O	3N		77ζ
=5=	–C+N–I+M+H	4C	4H	1O	3N		79ζ
=5=	+G–P–L+R+Y	4C	4H	1O	3N		92ζ
=5=	–P–L+R+Y–*	4C	3H	1O	3N	–1*	92ζ
=5=	+G–P–I+R+Y	4C	4H	1O	3N		94ζ
=5=	–P–I+R+Y–*	4C	3H	1O	3N	–1*	94ζ
=6=	–G+P+N–L+H–ϕ	5C	4H	1O	3N		–1ϕ 45ζ
=6=	+P+N–L+H+*–ϕ	5C	5H	1O	3N	1*	–1ϕ 45ζ
=6=	–G+P+N–I+H–ϕ	5C	4H	1O	3N		–1ϕ 47ζ
=6=	+P+N–I+H+*–ϕ	5C	5H	1O	3N	1*	–1ϕ 47ζ
=6=	+P–V+Q–K+R+ϕ	3C	4H	1O	3N		1ϕ 50ζ
=6=	–A+P–I+Q+H–ϕ	5C	4H	1O	3N		–1ϕ 51ζ
=6=	+P–T+D–L+R+ϕ	3C	4H	1O	3N		1ϕ 53ζ
=6=	–A+P–V+N+H–ϕ	5C	4H	1O	3N		–1ϕ 54ζ
=6=	+S–L+H–F+W+ϕ	3C	4H	1O	3N		1ϕ 54ζ

Contd.

Table 5.1 Contd

=6=	+G+A+N–L+H+ϕ	3C	4H	1O	3N		1ϕ 55ζ
=6=	+A+N–L+H–*+ϕ	3C	3H	1O	3N	–1*	1ϕ 55ζ
=6=	+P–T+D–I+R+ϕ	3C	4H	1O	3N		1ϕ 55ζ
=6=	+S–I+H–F+W+ϕ	3C	4H	1O	3N		1ϕ 56ζ
=6=	+G+A+N–I+H+ϕ	3C	4H	1O	3N		1ϕ 57ζ
=6=	–G+Q–K+R+*+ϕ	3C	5H	1O	3N	1*	1ϕ 57ζ
=6=	+A+N–I+H–*+ϕ	3C	3H	1O	3N	–1*	1ϕ 57ζ
=6=	+G–L+Q+H–*+ϕ	3C	3H	1O	3N	–1*	1ϕ 59ζ
=6=	+C–L–M+R+Y–ϕ	5C	4H	1O	3N		–1ϕ 60ζ
=6=	+T–V–L+K+H+ϕ	3C	4H	1O	3N		1ϕ 60ζ
=6=	–G–A+N+H+*–ϕ	5C	5H	1O	3N	1*	–1ϕ 61ζ
=6=	+G–I+Q+H–*+ϕ	3C	3H	1O	3N	–1*	1ϕ 61ζ
=6=	+C–I–M+R+Y–ϕ	5C	4H	1O	3N		–1ϕ 62ζ
=6=	+T–V–I+K+H+ϕ	3C	4H	1O	3N		1ϕ 62ζ
=6=	–P+D+Q–E+H+ϕ	3C	4H	1O	3N		1ϕ 63ζ
=6=	+G–V+N+H–*+ϕ	3C	3H	1O	3N	–1*	1ϕ 64ζ
=6=	–P+V–L+Q+H+ϕ	3C	4H	1O	3N		1ϕ 66ζ
=6=	–S–T+D+N+H–ϕ	5C	4H	1O	3N		–1ϕ 68ζ
=6=	–P+V–I+Q+H+ϕ	3C	4H	1O	3N		1ϕ 68ζ
=6=	–G–A–T+D+R+ϕ	3C	4H	1O	3N		1ϕ 69ζ
=6=	–A–T+D+R+*+ϕ	3C	5H	1O	3N	1*	1ϕ 69ζ
=6=	–P+N+I–L+H+ϕ	3C	4H	1O	3N		1ϕ 69ζ
=6=	+S–P–T+Q+H+ϕ	3C	4H	1O	3N		1ϕ 70ζ
=6=	+G–P+N+H+*+ϕ	3C	5H	1O	3N	1*	1ϕ 71ζ
=6=	–G–P+N+H–*+ϕ	3C	3H	1O	3N	–1*	1ϕ 71ζ
=6=	–P+N–I+L+H+ϕ	3C	4H	1O	3N		1ϕ 73ζ
=6=	+T–I–L+F+R–ϕ	5C	4H	1O	3N		–1ϕ 73ζ
=6=	+G–A–P+Q+H+ϕ	3C	4H	1O	3N		1ϕ 75ζ
=6=	–A–P+Q+H–*+ϕ	3C	3H	1O	3N	–1*	1ϕ 75ζ
=6=	+S–V–L+F+R–ϕ	5C	4H	1O	3N		–1ϕ 75ζ
=6=	–G–S–V+E+R+ϕ	3C	4H	1O	3N		1ϕ 77ζ
=6=	+S–V–I+F+R–ϕ	5C	4H	1O	3N		–1ϕ 77ζ
=6=	–S–V+E+R+*+ϕ	3C	5H	1O	3N	1*	1ϕ 77ζ
=6=	+A–V–L+R+Y–ϕ	5C	4H	1O	3N		–1ϕ 79ζ
=6=	+A–V–I+R+Y–ϕ	5C	4H	1O	3N		–1ϕ 81ζ
=6=	–G–P–V+R+Y–ϕ	5C	4H	1O	3N		–1ϕ 95ζ
=6=	–P–V+R+Y+*–ϕ	5C	5H	1O	3N	1*	–1ϕ 95ζ
		Guanine G78*					
=6=	–V+Q–K+H+R+ϕ	4C	4H	1O	5N		1ϕ 92ζ
=6=	–T+D–L+H+R+ϕ	4C	4H	1O	5N		1ϕ 95ζ

Contd.

Table 5.1 Contd

=6=	–T+D–I+H+R+ϕ	**4C**	**4H**	1**O**	5**N**		1ϕ 97ζ
=7=	–G–A–S–V+D+H+R	5**C**	4**H**	1**O**	5**N**		103ζ
=7=	–G–A–S–I+E+H+R	5**C**	4**H**	1**O**	5**N**		108ζ
=7=	–G–A–S–L+E+H+R	5**C**	4**H**	1**O**	5**N**		106ζ
=7=	–G–A–T–V+E+H+R	5**C**	4**H**	1**O**	5**N**		110ζ
=7=	–G–A+N–K+H+R+*	5**C**	5**H**	1**O**	5**N**	1*	91ζ
=7=	–G–S+D–I+H+R+*	5**C**	5**H**	1**O**	5**N**	1*	96ζ
=7=	–G–S+D–L+H+R+*	5**C**	5**H**	1**O**	5**N**	1*	94ζ
=7=	–G–P–V+N–I+R+W	5**C**	4**H**	1**O**	5**N**		116ζ
=7=	–G–P–V+N–L+R+W	5**C**	4**H**	1**O**	5**N**		114ζ
=7=	–G+P+N–I–K+H+R	5**C**	4**H**	1**O**	5**N**		77ζ
=7=	–G+P+N–L–K+H+R	5**C**	4**H**	1**O**	5**N**		75ζ
=7=	–G–P–I–L+Q+R+W	5**C**	4**H**	1**O**	5**N**		111ζ
=7=	–G–T–V+D+H+R+*	5**C**	5**H**	1**O**	5**N**	1*	98ζ
=7=	–G–T–I+E+H+R+*	5**C**	5**H**	1**O**	5**N**	1*	103ζ
=7=	–G–T–L+E+H+R+*	5**C**	5**H**	1**O**	5**N**	1*	101ζ
=7=	+G–V–I–L+Q+R+W	5**C**	4**H**	1**O**	5**N**		104ζ
=7=	–A–S–V+D+H+R+*	5**C**	5**H**	1**O**	5**N**	1*	103ζ
=7=	–A–S–I+E+H+R+*	5**C**	5**H**	1**O**	5**N**	1*	108ζ
=7=	–A–S–L+E+H+R+*	5**C**	5**H**	1**O**	5**N**	1*	106ζ
=7=	–A+P–V+N–K+H+R	5**C**	4**H**	1**O**	5**N**		84ζ
=7=	–A–P–V–I+Q+R+W	5**C**	4**H**	1**O**	5**N**		120ζ
=7=	–A–P–V–L+Q+R+W	5**C**	4**H**	1**O**	5**N**		118ζ
=7=	–A+P–I+Q–K+H+R	5**C**	4**H**	1**O**	5**N**		81ζ
=7=	–A+P–L+Q–K+H+R	5**C**	4**H**	1**O**	5**N**		79ζ
=7=	–A–T–V+E+H+R+*	5**C**	5**H**	1**O**	5**N**	1*	110ζ
=7=	+A–V+N–I–L+R+W	5**C**	4**H**	1**O**	5**N**		100ζ
=7=	–S+P–V+D–I+H+R	5**C**	4**H**	1**O**	5**N**		89ζ
=7=	–S+P–V+D–L+H+R	5**C**	4**H**	1**O**	5**N**		87ζ
=7=	–S+P–I–L+E+H+R	5**C**	4**H**	1**O**	5**N**		92ζ
=7=	–S–T+D+N–K+H+R	5**C**	4**H**	1**O**	5**N**		98ζ
=7=	+P–T–V–I+E+H+R	5**C**	4**H**	1**O**	5**N**		96ζ
=7=	+P–T–V–L+E+H+R	5**C**	4**H**	1**O**	5**N**		94ζ
=7=	–P–V+N–I+R+W+*	5**C**	5**H**	1**O**	5**N**	1*	116ζ
=7=	–P–V+N–L+R+W+*	5**C**	5**H**	1**O**	5**N**	1*	114ζ
=7=	+P+N–I–K+H+R+*	5**C**	5**H**	1**O**	5**N**	1*	77ζ
=7=	+P+N–L–K+H+R+*	5**C**	5**H**	1**O**	5**N**	1*	75ζ
=7=	–P–I–L+Q+R+W+*	5**C**	5**H**	1**O**	5**N**	1*	111ζ
=7=	+C+N–I–L–M+R+W	5**C**	4**H**	1**O**	5**N**		81ζ
=7=	–V–I–L+Q+R+W–*	5**C**	3**H**	1**O**	5**N**	–1*	104ζ

Contd.

Table 5.1 Contd

=7=	+N–I–L+Q–E+R+Y	5**C**	4**H**	1**O**	5**N**			94ζ
	^{6}N-Methyladenine A78*							
=6=	–T+Q–K+H+R+φ	5**C**	6**H**		5**N**		1φ	97ζ
=6=	–G–V+H+R+*+φ	5**C**	7**H**		5**N**	1*	1φ	98ζ
=6=	–S+N–K+H+R+φ	5**C**	6**H**		5**N**		1φ	98ζ
=7=	–G–A–S–T+D+H+R	6**C**	6**H**		5**N**			108ζ
=7=	–G–A+P–I+H+R+*	6**C**	7**H**		5**N**	1*		87ζ
=7=	–G–A+P–L+H+R+*	6**C**	7**H**		5**N**	1*		85ζ
=7=	+G–S+N–I–L+R+W	6**C**	6**H**		5**N**			110ζ
=7=	–G–S+Q–K+H+R+*	6**C**	7**H**		5**N**	1*		96ζ
=7=	–G–P–T+N–I+R+W	6**C**	6**H**		5**N**			121ζ
=7=	–G–P–T+N–L+R+W	6**C**	6**H**		5**N**			119ζ
=7=	+G–T–V+N–I+R+W	6**C**	6**H**		5**N**			114ζ
=7=	+G–T–V+N–L+R+W	6**C**	6**H**		5**N**			112ζ
=7=	+G–T–I–L+Q+R+W	6**C**	6**H**		5**N**			109ζ
=7=	–G–V–I–L+K+R+W	6**C**	6**H**		5**N**			110ζ
=7=	–A–S–P+N–I+R+W	6**C**	6**H**		5**N**			126ζ
=7=	–A–S–P+N–L+R+W	6**C**	6**H**		5**N**			124ζ
=7=	–A–S–T+D+H+R+*	6**C**	7**H**		5**N**	1*		108ζ
=7=	–A–P–T–V+N+R+W	6**C**	6**H**		5**N**			128ζ
=7=	–A+P–T+N–K+H+R	6**C**	6**H**		5**N**			89ζ
=7=	–A–P–T–I+Q+R+W	6**C**	6**H**		5**N**			125ζ
=7=	–A–P–T–L+Q+R+W	6**C**	6**H**		5**N**			123ζ
=7=	+A–T+N–I–L+R+W	6**C**	6**H**		5**N**			105ζ
=7=	–S+P–T+D–I+H+R	6**C**	6**H**		5**N**			94ζ
=7=	–S+P–T+D–L+H+R	6**C**	6**H**		5**N**			92ζ
=7=	–S+P–V+Q–K+H+R	6**C**	6**H**		5**N**			89ζ
=7=	+S–P–I–E+H+R+Y	6**C**	6**H**		5**N**			115ζ
=7=	+S–P–L–E+H+R+Y	6**C**	6**H**		5**N**			113ζ
=7=	–S+N–I–L+R+W–*	6**C**	5**H**		5**N**	–1*		110ζ
=7=	–P–T+N–I+R+W+*	6**C**	7**H**		5**N**	1*		121ζ
=7=	–P–T+N–L+R+W+*	6**C**	7**H**		5**N**	1*		119ζ
=7=	–T–V+N–I+R+W–*	6**C**	5**H**		5**N**	–1*		114ζ
=7=	–T–V+N–L+R+W–*	6**C**	5**H**		5**N**	–1*		112ζ
=7=	–T–I–L+Q+R+W–*	6**C**	5**H**		5**N**	–1*		109ζ
=7=	–V+N–I+Q–E+F+R	6**C**	6**H**		5**N**			100ζ
=7=	–V+N–L+Q–E+F+R	6**C**	6**H**		5**N**			98ζ
=7=	–V–I–L+K+R+W+*	6**C**	7**H**		5**N**	1*		110ζ
=7=	–D+N–I–L+Q+F+R	6**C**	6**H**		5**N**			103ζ

Contd.

Table 5.1 Contd

		5-Methylcytosine C66*					
=5=	-G+C+Q-M+H	5**C**	6**H**	1**O**	3**N**		45ζ
=5=	+C+Q–M+H+*	5**C**	7**H**	1**O**	3**N**	1*	45ζ
=5=	+G+P+N–L+H	5**C**	6**H**	1**O**	3**N**		53ζ
=5=	+P+N–L+H–*	5**C**	5**H**	1**O**	3**N**	–1*	53ζ
=5=	+G+P+N–I+H	5**C**	6**H**	1**O**	3**N**		55ζ
=5=	+P+N–I+H–*	5**C**	5**H**	1**O**	3**N**	–1*	55ζ
=5=	–G+D+N–E+H	5**C**	6**H**	1**O**	3**N**		57ζ
=5=	+D+N–E+H+*	5**C**	7**H**	1**O**	3**N**	1*	57ζ
=5=	–G+V+N–L+H	5**C**	6**H**	1**O**	3**N**		60ζ
=5=	+V+N–L+H+*	5**C**	7**H**	1**O**	3**N**	1*	60ζ
=5=	–A+D+Q–E+H	5**C**	6**H**	1**O**	3**N**		61ζ
=5=	–G+V+N–I+H	5**C**	6**H**	1**O**	3**N**		62ζ
=5=	+V+N–I+H+*	5**C**	7**H**	1**O**	3**N**	1*	62ζ
=5=	+Q–K+H–F+W	5**C**	6**H**	1**O**	3**N**		63ζ
=5=	–G+A–V+Q+H	5**C**	6**H**	1**O**	3**N**		64ζ
=5=	–G+S–T+N+H	5**C**	6**H**	1**O**	3**N**		64ζ
=5=	–A+V–L+Q+H	5**C**	6**H**	1**O**	3**N**		64ζ
=5=	+A–V+Q+H+*	5**C**	7**H**	1**O**	3**N**	1*	64ζ
=5=	+S–T+N+H+*	5**C**	7**H**	1**O**	3**N**	1*	64ζ
=5=	–A+V–I+Q+H	5**C**	6**H**	1**O**	3**N**		66ζ
=5=	–A+N+I–L+H	5**C**	6**H**	1**O**	3**N**		67ζ
=5=	–V+D+H–Y+W	5**C**	6**H**	1**O**	3**N**		67ζ
=5=	–A+S–T+Q+H	5**C**	6**H**	1**O**	3**N**		68ζ
=5=	+G–A+N+H+*	5**C**	7**H**	1**O**	3**N**	1*	69ζ
=5=	–G–A+N+H–*	5**C**	5**H**	1**O**	3**N**	–1*	69ζ
=5=	–S+T–V+Q+H	5**C**	6**H**	1**O**	3**N**		69ζ
=5=	–L+E+H–Y+W	5**C**	6**H**	1**O**	3**N**		70ζ
=5=	–A+N–I+L+H	5**C**	6**H**	1**O**	3**N**		71ζ
=5=	–S+D–L+K+H	5**C**	6**H**	1**O**	3**N**		72ζ
=5=	–T+Q+H–F+Y	5**C**	6**H**	1**O**	3**N**		72ζ
=5=	–I+E+H–Y+W	5**C**	6**H**	1**O**	3**N**		72ζ
=5=	–S+N+H–F+Y	5**C**	6**H**	1**O**	3**N**		73ζ
=5=	–S+D–I+K+H	5**C**	6**H**	1**O**	3**N**		74ζ
=5=	–T–V+D+K+H	5**C**	6**H**	1**O**	3**N**		76ζ
=5=	–V–D+Q+E+H	5**C**	6**H**	1**O**	3**N**		76ζ
=5=	–T–L+E+K+H	5**C**	6**H**	1**O**	3**N**		79ζ
=5=	–T–I+E+K+H	5**C**	6**H**	1**O**	3**N**		81ζ

Contd.

Table 5.1 Contd

=5=	–N+E–K+H+R	5**C**	6**H**	1**O**	3**N**		81ζ
=5=	–C–L+Q+M+H	5**C**	6**H**	1**O**	3**N**		83ζ
=5=	–C–I+Q+M+H	5**C**	6**H**	1**O**	3**N**		85ζ
=5=	–C–V+N+M+H	5**C**	6**H**	1**O**	3**N**		88ζ
=5=	+S–P–L+F+R	5**C**	6**H**	1**O**	3**N**		90ζ
=5=	+S–P–I+F+R	5**C**	6**H**	1**O**	3**N**		92ζ
=5=	+A–P–L+R+Y	5**C**	6**H**	1**O**	3**N**		94ζ
=5=	+A–P–I+R+Y	5**C**	6**H**	1**O**	3**N**		96ζ
=5=	+G–P–V+R+Y	5**C**	6**H**	1**O**	3**N**		103ζ
=5=	–P–V+R+Y–*	5**C**	5**H**	1**O**	3**N**	–1*	103ζ
=5=	+G–P–V+R+Y	5**C**	6**H**	1**O**	3**N**		103ζ
=5=	–P–V+R+Y–*	5**C**	5**H**	1**O**	3**N**	–1*	103ζ
=6=	+G+C+N–M+H+ϕ	4**C**	6**H**	1**O**	3**N**		1ϕ 47ζ
=6=	+C+N–M+H–*+ϕ	4**C**	5**H**	1**O**	3**N**	–1*	1ϕ 47ζ
=6=	–G+P–L+Q+H–ϕ	6**C**	6**H**	1**O**	3**N**		–1ϕ 51ζ
=6=	+P–L+Q+H+*–ϕ	6**C**	7**H**	1**O**	3**N**	1*	–1ϕ 51ζ
=6=	–G+P–I+Q+H–ϕ	6**C**	6**H**	1**O**	3**N**		–1ϕ 53ζ
=6=	+P–I+Q+H+*–ϕ	6**C**	7**H**	1**O**	3**N**	1*	–1ϕ 53ζ
=6=	–G+P–V+N+H–ϕ	6**C**	6**H**	1**O**	3**N**		–1ϕ 56ζ
=6=	+P–V+N+H+*–ϕ	6**C**	7**H**	1**O**	3**N**	1*	–1ϕ 56ζ
=6=	–A+P+N–K+R+ϕ	4**C**	6**H**	1**O**	3**N**		1ϕ 57ζ
=6=	–A+P–V+Q+H–ϕ	6**C**	6**H**	1**O**	3**N**		–1ϕ 60ζ
=6=	–S+P+D–L+R+ϕ	4**C**	6**H**	1**O**	3**N**		1ϕ 60ζ
=6=	+G+A–L+Q+H+ϕ	4**C**	6**H**	1**O**	3**N**		1ϕ 61ζ
=6=	+A–L+Q+H–*+ϕ	4**C**	5**H**	1**O**	3**N**	–1*	1ϕ 61ζ
=6=	+T–L+H–F+W+ϕ	4**C**	6**H**	1**O**	3**N**		1ϕ 61ζ
=6=	–S+P+D–I+R+ϕ	4**C**	6**H**	1**O**	3**N**		1ϕ 62ζ
=6=	+G+A–I+Q+H+ϕ	4**C**	6**H**	1**O**	3**N**		1ϕ 63ζ
=6=	+A–I+Q+H–*+ϕ	4**C**	5**H**	1**O**	3**N**	–1*	1ϕ 63ζ
=6=	+T–I+H–F+W+ϕ	4**C**	6**H**	1**O**	3**N**		1ϕ 63ζ
=6=	+P–T–V+D+R+ϕ	4**C**	6**H**	1**O**	3**N**		1ϕ 64ζ
=6=	+S–V+H–F+W+ϕ	4**C**	6**H**	1**O**	3**N**		1ϕ 65ζ
=6=	+G+A–V+N+H+ϕ	4**C**	6**H**	1**O**	3**N**		1ϕ 66ζ
=6=	–A+S–L+K+H+ϕ	4**C**	6**H**	1**O**	3**N**		1ϕ 66ζ
=6=	+A–V+N+H–*+ϕ	4**C**	5**H**	1**O**	3**N**	–1*	1ϕ 66ζ
=6=	–G–A+Q+H+*–ϕ	6**C**	7**H**	1**O**	3**N**	1*	–1ϕ 67ζ

Contd.

Table 5.1 Contd

=6=	+P–T–L+E+R+ϕ	4C	6H	1O	3N		1ϕ 67ζ
=6=	–A+S–I+K+H+ϕ	4C	6H	1O	3N		1ϕ 68ζ
=6=	+P–T–I+E+R+ϕ	4C	6H	1O	3N		1ϕ 69ζ
=6=	+G–V+Q+H–*+ϕ	4C	5H	1O	3N	–1*	1ϕ 70ζ
=6=	–L+K+H–F+Y+ϕ	4C	6H	1O	3N		1ϕ 70ζ
=6=	–G–T+D+R+*+ϕ	4C	7H	1O	3N	1*	1ϕ 71ζ
=6=	+C–V–M+R+Y–ϕ	6C	6H	1O	3N		–1ϕ 71ζ
=6=	–I+K+H–F+Y+ϕ	4C	6H	1O	3N		1ϕ 72ζ
=6=	–G+A–P+N+H+ϕ	4C	6H	1O	3N		1ϕ 73ζ
=6=	+A–P+N+H+*+ϕ	4C	7H	1O	3N	1*	1ϕ 73ζ
=6=	–S–T+D+Q+H–ϕ	6C	6H	1O	3N		–1ϕ 74ζ
=6=	–P+I–L+Q+H+ϕ	4C	6H	1O	3N		1ϕ 75ζ
=6=	–G–A–S+D+R+ϕ	4C	6H	1O	3N		1ϕ 76ζ
=6=	–A–S+D+R+*+ϕ	4C	7H	1O	3N	1*	1ϕ 76ζ
=6=	+S–L–K+R+W–ϕ	6C	6H	1O	3N		–1ϕ 76ζ
=6=	+G–P+Q+H+*+ϕ	4C	7H	1O	3N	1*	1ϕ 77ζ
=6=	–G–P+Q+H–*+ϕ	4C	5H	1O	3N	–1*	1ϕ 77ζ
=6=	–S–P+T+N+H+ϕ	4C	6H	1O	3N		1ϕ 78ζ
=6=	+S–I–K+R+W–ϕ	6C	6H	1O	3N		–1ϕ 78ζ
=6=	+D–L–E+R+Y–ϕ	6C	6H	1O	3N		–1ϕ 78ζ
=6=	–P–I+L+Q+H+ϕ	4C	6H	1O	3N		1ϕ 79ζ
=6=	–P–V+N+I+H+ϕ	4C	6H	1O	3N		1ϕ 80ζ
=6=	+D–I–E+R+Y–ϕ	6C	6H	1O	3N		–1ϕ 80ζ
=6=	–S–T+N+E+H–ϕ	6C	6H	1O	3N		–1ϕ 82ζ
=6=	–P–V+N+L+H+ϕ	4C	6H	1O	3N		1ϕ 82ζ
=6=	+T–V–L+F+R–ϕ	6C	6H	1O	3N		–1ϕ 82ζ
=6=	–G–A–T+E+R+ϕ	4C	6H	1O	3N		1ϕ 83ζ
=6=	–A–T+E+R+*+ϕ	4C	7H	1O	3N	1*	1ϕ 83ζ
=6=	+V–I–L+R+Y–ϕ	6C	6H	1O	3N		–1ϕ 83ζ
=6=	+T–V–I+F+R–ϕ	6C	6H	1O	3N		–1ϕ 84ζ
=6=	+S–T–L+R+Y–ϕ	6C	6H	1O	3N		–1ϕ 85ζ
=6=	–P–D+N+E+H+ϕ	4C	6H	1O	3N		1ϕ 85ζ
=6=	+N–L–Q+R+Y–ϕ	6C	6H	1O	3N		–1ϕ 86ζ
=6=	+S–T–I+R+Y–ϕ	6C	6H	1O	3N		–1ϕ 87ζ
=6=	+N–I–Q+R+Y–ϕ	6C	6H	1O	3N		–1ϕ 88ζ
=6=	+G–A–L+R+Y–ϕ	6C	6H	1O	3N		–1ϕ 90ζ
=6=	–A–L+R+Y–*–ϕ	6C	5H	1O	3N	–1*	–1ϕ 90ζ

Contd.

Table 5.1 Contd

=6=	+G–A–I+R+Y–ϕ	6**C**	6**H**	1**O**	3**N**		-1ϕ 92ζ
=6=	–A–I+R+Y–*–ϕ	6**C**	5**H**	1**O**	3**N**	–1*	-1ϕ 92ζ
=6=	–P+N–K+F+R–ϕ	6**C**	6**H**	1**O**	3**N**		-1ϕ 93ζ

Dominant charge configurations of DNA and RNA nucleotides

The state $U46_\phi = Q_p(29_{-1^*}) = -1\mathbf{C} -1\mathbf{H} +2\mathbf{O} +3\mathbf{N}$ is a CFAA basis with metric 2^+2^- conjugated with protein basis. Vector $U46_\phi$ shifts the fields **GF**(137) on 30 units; therefore, the pair **AT** induces the field **GF**(167). Appearance of the metric 2^+2^- in an internal charge space of DNA leads to the extrusion of Coulomb field to the DNA surface, i.e., to the hypochromic effect. Moreover, the non-integrated part of informational nucleotide sequence, appears in DNA. This part is called intron. And indeed, the Fourier spectrum of DNA cannot give exact answer about its structure.

Let us examine the circulation of CFAA vector around a closed loop (Fig. 5.1). The existence of an ideal (I, L) with a commutator –2 determines the function Sgn(ϕ) of the field of vectors **GF**(3) parallel transfer.

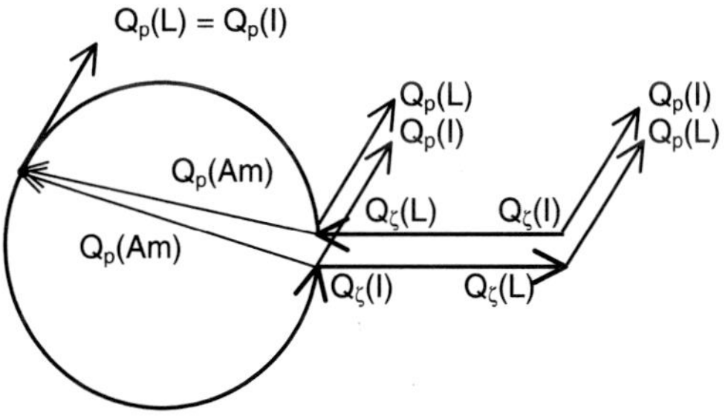

Fig. 5.1 Loop 'C' of the ideal (I, L) expansion in the field ϕ with a spin of 5/2. Circulation of vector Am along the loop leads to the different values of Q_p(Am); however, tangent vectors coincide $Q_p(I) = Q_p(L)$. The marked point lies on the loop itself and not inside of it. Deviation of vectors Am in the points Q_ζ(L) and Q_ζ(I) is caused by existence of 4_ϕ-configurations of Am. Vectors Q_ζ(I, L) and Q_ζ(L, I) are antisymmetric in the joint points. Ideal (I, L) extends to the paired complementary nucleotides of RNA, the loop 'C' forms the unpaired part of RNA nucleotides. The enzyme, whose metric tensor coincides with its geometric tensor is the marked point of RNA.

A point $Q_p(I) = Q_p(L)$ can be found on contour 'C'. We can say that **whatever is the measurement mode of the parameters of the genetic code amino acids, the parameters of two from 20 amino acids will always coincide completely, because it is impossible to remove the marked point of a terminator by any transformation of CFAA vectors.**

For the ideal (I, L), the identity $Q_\zeta(I) \neq Q_\zeta(L)$ is true. All cases of $\delta\phi$ and $Q_{p\zeta}$ (Am) are encoded in DNA. The loop 'C' itself is a continuation of inequality $Q_\zeta(I) \neq Q_\zeta(L)$.

Gauge invariancy of ordinal numbers **Q1, Q2, Q3, Q4** is in fact that the CFAA vectors can be selected in an arbitrary frame, but such vectors will always be found, for which **Q1** = **Q4** (identity of global conservation of marked point) and **Q2** = **Q3** (identity of local breaking of marked point).

Configurations of amino acids, which have 3 and 4_ϕ indexes determine an external covariant derivative of protein, whereas the field A_V. A1, A2, A3_A2', A3' determines the proper derivative of the protein. It is especially important for the enzyme, because components of the field A1, A2, A3, which have proper charges $Q_4(A1)$, $Q_4(A2)$, $Q_4(A3)$ preserve the metric tensor e_1, e_2, e_3 of RNA(N1, N2, N3):

$$e_1 = h(A1);\ e_2 = h(A2);\ e_3 = h(A3).$$

Field ϕ is built in such a way that three-index configurations of amino acids that are included in protein composition have the same RNA metric tensor:

$$A_V = + e_1 Q_p(Am1) + e_2 Q_p(Am2) + e_3 Q_p(Am3).$$

Amino acids in protein obey Fermi statistics and can be arranged analogically to the genetic code amino acids (Chapter 4) in the local field **GF**(Am). Then all amino acids will have non-coincident charges and 20-parametric Lie group will be the protein movement group. The field **GF**(Am) determines the bundle of protein over the base $Q_p\{Am + 29\}$ of paired amino acids with the basic amino acid Q(29) of a protein backbone.

If the bundled field GF(Am) is non-trivial, i.e., is not equal to GF(29), then protein is in an ordered state; if GF(Am) = GF(29) then amino acids obey Bose statistics and protein folds into a statistical tangle.

Let us examine the involutive transformation DNA $\leftrightarrow$ RNA on the basis of 3- and 4_ϕ-configurations of basic thymine $T55_{-2}\{9^+\ 3^-\}$ and basic uracil $U45_{-1*}\{8^+\ 1^-\}$. Informational nucleotide **T66*** has metric $1^+\ 1^-$ in a four-dimensional charge space of RNA with metric $2^+\ 2^-$. Metric tensors of basic thymine and basic uracil divide **T66*** into three parts:

1. Metric tensor of CFAA

$$T55_{-2}\{9^+3^-\} \,/\, U45_{-1^*}\{8^+1^-\} = (9^+ - 3^-, 8^+ + 1^-)_{-1^*} = 13^+7^-.$$

2. **T66*** – marked point of DNA

$$T55_{-2}\{9^+3^-\} \,//\, U45_{-1^*}\{8^+1^-\} = (9^+ - 1^-, -8^+ + 3^-)^* = 10^+10^-.$$

induces reverse transformation of RNA into DNA.

3. Metric tensor of RNA

$$T55_{-2}\{9^+3^-\} \,///\, U45_{-1^*}\{8^+1^-\} = (-3^-, 1^-) = 2^+2^-.$$

In finite projected plane P(4), the following identity is fulfilled:

$$P(4)\{13^+7^-\} + P(4)^{+2}\{10^+10^-\} + \phi\{19^+2^-\} + \mathbf{GF}(5^*)\,\{2^+2^-\} = \mathbf{T66^*}.$$

Metric tensor of CFAA charge space is a representation of division mode of interval (–1, +1).

The unity of proton charge in the field **GF**(83) is higher than in the field W = **GF**(73), but due to integer value of amino acid proton charge when compressing $^{\mathbf{GF}(83)} \downarrow _{\mathbf{GF}(73)}$ modulo10-parametric Poincaré group, proton charges of some amino acids inevitably have to coincide. This leads to the appearance of CFAA marked points, which preserve **GF**(83) field. Charge of Galois field for ordinal type ω_2 is incidental to the straight-line segment; therefore, in true gravitational field that contains the contraction point, appearance of equivalent charge configurations of amino acids is inevitable.

The set of equivalent charge configurations is closely related by its meaning to the wave function of particle in quantum mechanics. Irreversibility of compression process of the field **GF**(83) and the absence of CFAA collapse is possible only in case if CFAA contains marked point of terminator, i.e., with any CFAA compression mode, the parameters of two from 20 amino acids will be completely identical. Otherwise, transfer **GF**(83) ↔ **GF**(73) and, hence, transformation of DNA into RNA will be irreversible.

The marked points of CFAA (contact points of attraction) form a region of soft contact interaction of biological molecules, which are asymptotically free. Asymptotic freedom is typical for the phase III of CFAA doubling process, which is presented in Fig. 2.5. It is a consequence of the bundle of CFAA doubled Riemann space.

Euclidean space of informational nucleotides

Informational nucleotides **N**inf with the marked point are determined in external Euclidean space, since

$$h(\mathbf{A}70^*) = h(\mathbf{T}66^*) = h(\mathbf{G}78^*) = h(\mathbf{C}58^*) = h(\mathbf{U}58^*) = +1,$$

where the function h(**N**inf) = Q_p(**N**inf) (mod 4). Therefore, triplets of RNA nucleotides are formed as linear combinations of nucleotides but with hyperbolic metric of terminators' space.

Let us introduce the symbol $\theta(N_3$*inf) = Q_p{+N1+N2–N3} of line connection of RNA informational triplets with the marked point in the third triplet position and let us project RNA triplets onto the amino acids of the genetic code.

Minimum value of $\theta(N_3$*inf) is transferred onto the amino acids in the centre of CFAA with the aid of field **GF**(3):

$$\theta(N_3^*\mathrm{inf})\mathrm{min} = Q_p\{\mathbf{U}58 + \mathbf{U}58 - \mathbf{G}78\} + \mathbf{GF}(3) = M(41).$$

Maximum value of $\theta(N_3$*inf) is transferred on amino acids with the aid of field **GF**(5) and if we take into account the covariant charge equal to 20 marked points in the doubled space **GF**(167) of finite projected plane P(4), then:

$$\theta(N_3^*\mathrm{inf})\mathrm{max} = Q_p\{\mathbf{G}78 + \mathbf{G}78 - \mathbf{U}58\} - \mathbf{GF}(5) - P(4)^* = W(73).$$

And, finally, let us take into account the reflection onto 32* with regard to the centre M(41) of terminators' space: A(9*) = M(41)* –32. Obtained marked points A, M, W will be reflected on points of interval 64** of the genetic code amino acids. The arranged distribution of the genetic code amino acids obtained at the end of Chapter 4 is the result of the projection.

C, N - Protein states

Let us examine in more detail the trigger cell of protein with **O**-input Q29(O) and **N**-input Q'29(N) (Fig. 3.7).

Let us introduce the marked point $\mathbf{CH_2}$, which determines the connection of amino acid radical to **C**-output of the bound trigger pair Q(29), Q'29:

$$Q_4(\mathbf{n}) = \mathbf{H_2O/CH_4} = (\mathbf{CH_2})^*\mathbf{O}.$$

Then the connection of amino acid radical to **N**-input symmetrizes the trigger Q(29). If the marked point is located on the **C**-end of the trigger, then it determines the local **C**-state of the protein; but if the marked point is taken away through **O**-input of trigger by exchange proton, the local state of protein is determined as **N**-state.

The marked point $\mathbf{CH_2}$ differentiates the states of trigger pair and determines the amino acid radical, which is dominant during the forming of three-dimensional protein structure.

All the radicals of amino acids, except glycine, contain the marked point $\mathbf{CH_2}$. If glycine is connected to the **C**-output of the trigger pair, then the marked point $\mathbf{CH_2}$ is located on the base Q(29), because $Q_p(\mathbf{n}) = \mathbf{NH}/\mathbf{CH_2}$.

Let us introduce the operators $\mathbf{N^+\{CH_2/H\}}$ and $\mathbf{N^-\{N/H\}}$ of the exchanging marked point $Q_p(\mathbf{n})$ between the triggers Q29(**O**) and Q′29(**N**):

$$N^+Q29(\mathbf{O}) = Q29/\mathbf{O} = \begin{cases} \mathbf{CH_2} \\ \mathbf{C}; \\ \mathbf{N} \end{cases} \quad N^-Q'29(\mathbf{N}) = Q'29/2C = \begin{cases} \mathbf{NH} \\ \mathbf{H} \\ \mathbf{CH_2} \end{cases}$$

with a commutator equal to

$$Q_p[\{2C\,2H\,1N = P(4)\} - \{1C\,4H\,1N = S^{-*}\}] = +4.$$

It means that the dimension of the finite projected plane P(4) coincides with the dimension of the projection space of triggers Q29(**O**), Q′29(**N**) states.

It follows from the structure of trigger pair states that the input **O** is a clock input and input **N** is an informational input of the trigger pair. States Q29(**C**) and Q′29(**H**) are the indefinite, metastable states.

Switching of trigger pair Q29(O), Q′29(N) with the exchanging marked point Q_p(n) determines the proper moment of protein movement quantity and its biological activity.

Normal protein structure is determined as a metastable **C**-state of protein and is characterized by helical protein zones; **N**-state of protein is stable. A local metastable **N**-state of protein distinguishes the dominant amino acid G and corresponds to the linear protein sites.

If two or more radicals of amino acids, which are connected to the N-terminus of a trigger pair, are present in the protein, then such a protein can be turned into the metastable local N-state. This situation is typical for prions and folded protein structures.

Let us reflect the state of field ϕ in a mixed gauge onto the states of triggers Q′29(**N**), Q29(**O**) (Fig. 5.2)

The marked point $\mathbf{CH_2}$ is included in space of states of both triggers, therefore, in phase I, let us reflect the state of field ϕ directly onto the trigger states. Let us extract the set of nucleotides Am(3) and compare it with the set of chemical elements designating the trigger states:

$$Am(3) = \{N1N1\; N2N2N2N2^*\; N3N3N3\}$$

$$Q29(\mathbf{O}), Q'29(\mathbf{N}) = \{\mathbf{NN\; H2H2HH\; CCC}\}.$$

From here we obtain the first gauge of states of field ϕ:

$$\phi^{-2} = \text{N1}(\mathbf{N})\,\text{N2}(\mathbf{H}_2)\,\text{N3}(\mathbf{C}).$$

State ϕ^{-2} is fixed by transition N2* $\rightarrow$ N2. This transition induces transition N2N2N3 $\rightarrow$ N2N3N3 and, consequently, transition N2* $\rightarrow$ N3 = **H** $\rightarrow$ **C**, which determines the metric 5^+1^- of field ϕ^{-2}.

In phase II field ϕ has the form

$$\phi^{-1} = \begin{vmatrix} \text{N1N2N3} & \text{N2N1N3}^* \\ \text{N1N3N2} & \text{N2N3N1} \\ \text{N2N3N2}^* & \text{N3N1N3}^* \end{vmatrix}$$

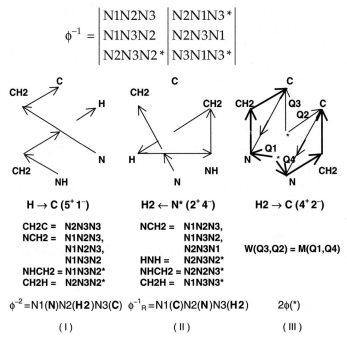

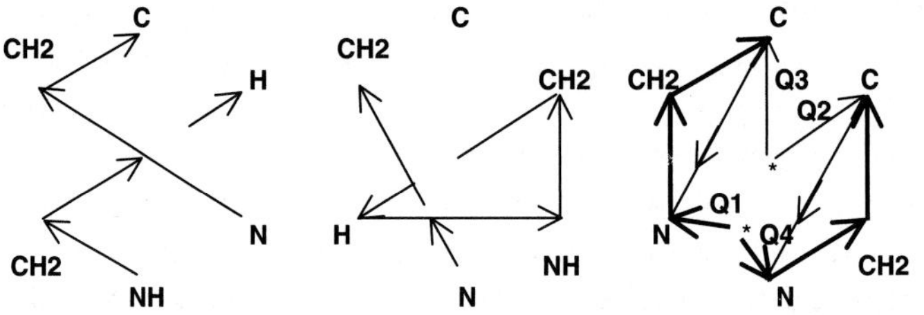

H → C (5^+1^-) **H2 ← N* (2^+4^-)** **H2 → C (4^+2^-)**

CH2C = N2N3N3	**NCH2 = N1N2N3,**	
NCH2 = N1N2N3,	**N1N3N2,**	
N1N2N3,	**N2N3N1**	**W(Q3,Q2) = M(Q1,Q4)**
N1N3N2	**HNH = N2N3N2***	
NHCH2 = N1N3N2*	**NHCH2 = N2N2N3***	
CH2H = N2N3N2*	**CH2H = N1N3N3***	

ϕ^{-2}=N1(**N**)N2(**H2**)N3(**C**) $\phi^{-1}{}_{R}$=N1(**C**)N2(**N**)N3(**H2**) 2ϕ(*)

(I) (II) (III)

Fig. 5.2 Duplication of conformal field onto the states of trigger Q' 29(**N**)$_{I}$, Q29(**O**)$_{II}$. Field of type W is highlighted; its copy is a field of type M(41) = Q_p(**CCH2N*NCH2C**).

Left field ϕ^{-1} remained in former gauge, whereas right field $\phi^{-1}{}_{R}$ is in a new gauge

$$\phi^{-1}{}_{R} = \text{N1}(\mathbf{C})\,\text{N2}(\mathbf{N})\,\text{N3}(\mathbf{H}_2).$$

Transition N3N2* $\leftarrow$ N3N3 or N2* $\leftarrow$ N3 = $\mathbf{H}_2$ $\leftarrow$ **N*** induces the gauge $\phi^{-1}{}_{R}$ and determines the metric 2^+4^- of field ϕ^{-1}. Phases I and II are the bundle of identical transition N3N2*N3 $\leftrightarrow$ N3N2*N3.

In phase III, field ϕ is in the duplication stage, the metric of field 4^+2^- is completely internal and is determined by the marked points of identical transition.

In phase I, proton binds two planes of parallel transfer of field ϕ, because it is not possible to shield conformal field of amino acids.

The conformal field of amino acids contracts the unmarked states of amino acids into one vector; the marked states of amino acids switch the trigger pair Q29(O), Q′29(N) and, hence, change the proper moment of protein movement quantity.

The state of trigger Q29(**O**) is related to phase II of clock transition of **C**, **N** – trigger when connecting the radical of amino acid Am to tRNA in three gauges N1N2N3, N1N3N2 and N2N3N1.

The state of trigger Q′29(**N**) is related to phase I and II of CFAA informational fixation. Trigger Q′29(**N**) stores information about clock mRNA triplet in three gauges N1N2N3, N1N2N3 and N1N3N2, therefore, the trigger pair Q29(**O**), Q′29(**N**) has six states, which correspond to the maximum number of coding triplets of the amino acid Am.

Due to the absence of finite projected plane of order six, the previous gauge Am that corresponds to information Am(N1N3N2*) is dismissed and replaced by a new Am(N2N2N3) that determines the two remaining states of a trigger pair.

Thus, protein forms doubled Riemann space with basic amino acid Am(N1N2N3), which contains information about amino acid Am (N2 N3N3) and does not always coincide with previous information Am(N1N3N2*).

Cap-structure of matrix RNA

Analysis of mRNA cap-structure is based on building of equivalent charge configurations of RNA informational nucleotides. RNA methylation mark $\mathbf{CH_2}$ is used as an amplitude mark of wave movement of ribosome 40S subunit. Figure 5.3 shows how the starting marks of prokaryotes V(**GUG**), I(**AUA**) are being formed. Methylation of adenine is required to overcome the terminator *(**UGA**) and to replace it with the leader mark W(**UGG**). Copy of M(**AUG**) and mark V(**GUG**), I(**AUA**) are the bundle of terminator in a position A2 of field ϕ. With the mirror reflection of furanose ring with respect to the mark $\mathbf{CH_2}$, basic thymine turns into informational uracil.

Informational uracil U is the marked point of the copies of starting mark M(41):

$$\begin{aligned} V(\mathbf{GUG})_{CH2} &= M_{-4}(\mathbf{CH_2}); \\ I(\mathbf{AUA})_{CH2} &= M_{-3}(\mathbf{CH_2}); \\ M(\mathbf{AUG})_{CH2} &= M_{-2}(\mathbf{CH_2}). \end{aligned}$$

Thus, **the cap-structure is a universal starting sequence of RNA translation and contains all starting marks of prokaryotes.**

In order to describe the ribosome movement with respect to mRNA as well as tRNA movement during the process of translation, it is necessary to build representation of charge-exchanging group among mRNA, ribosome and tRNA.

Let us consider that the exchange of atom coordinates among these structures is determined by action of a group SU(2).

In a group SU(2), relative movement of structures 1 and 2 with its own spatial-temporal coordinates x, t is determined by relations:

$$\begin{cases} t_2 = +1/dt_1 + b/dx_2 \\ x_1 = -b/dt_1 + 1/dx_2; \end{cases} \tag{5.1}$$

$$\begin{cases} t_2 = +dt_1 + b\,x_1 \\ x_2 = +b\,t_1 + d\,x_1; \end{cases} \tag{5.2}$$

where $d = \sqrt{1+b^2}\,; b = \beta/\sqrt{1-\beta^2}$.

The velocity β of relative movement of the frames of structures 1 and 2 is determined with respect to the velocity of Coulomb field penetration into the volume of their interaction.

Equation (5.1) describes accompanying, mutual phase transformation of wave movement of the structures (wave function of ribosome is recognition of furanose ring).

Equation (5.2) is a Lorentz transformation.

The following relations are the ones of great importance:

$$\rho_1(\beta) = \beta/b = \sqrt{1-\beta^2}\,; \rho_2(b) = b/\beta = \sqrt{1+b^2}.$$

Terminator* mark for prokaryotes is directly connected to Shine-Dalgarno sequence and remains behind the ribosome movement; ρ-factor $\rho_1(\beta)$ contains only the basic marked point of field ϕ: $\rho_1(\beta) = \mathbf{C}/\mathbf{H_2O} = 3/5$.

On the contrary, the terminator mark in an eucaryotic mRNA cap-structure is an obstacle for the ribosome movement and ρ-factor $\rho_2(b)$ contains the marked point $\mathbf{CH_4}$ of factor metric of protein basis of field ϕ expansion: $\rho_2(b) = \mathbf{CH_4}/\mathbf{C} = 5/3$.

Mark $\mathbf{CH_2}$ of amino acids and basis +1**C** of field ϕ as an addition to cytosine **C**58 is a marked points for tRNA: $+1\mathbf{C} = Q_p\{\Psi(\mathrm{Am}) - \mathbf{C}58\}; b(\mathrm{tRNA}) = \mathbf{CH_2}/\mathbf{C} = 4/3$.

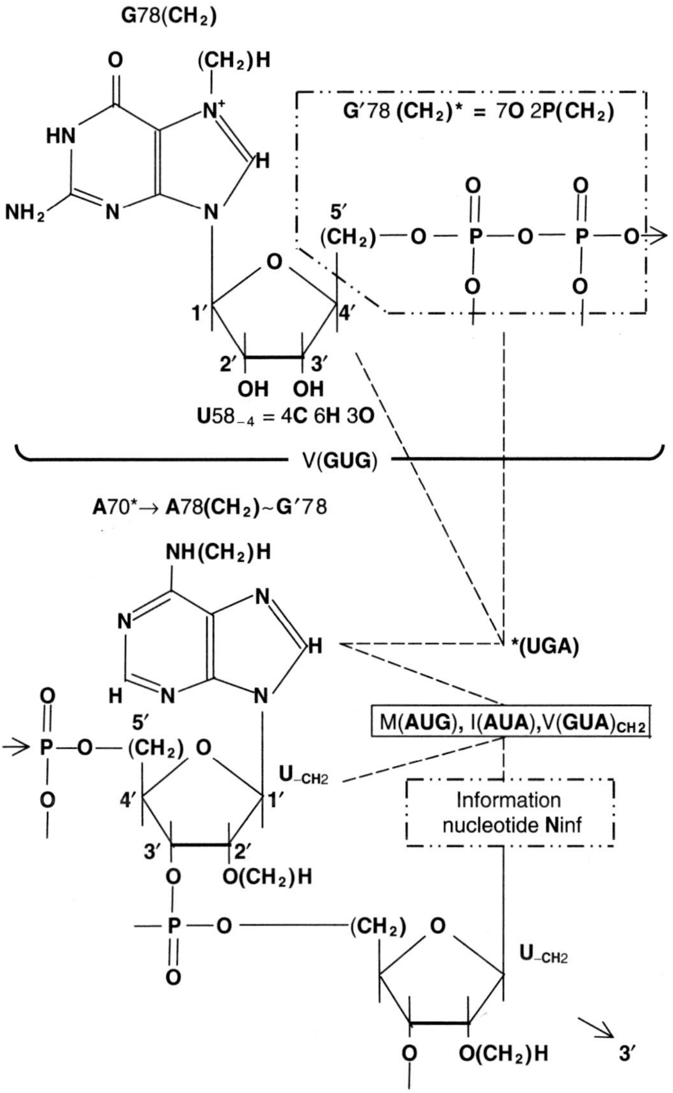

Fig. 5.3 Cap-structure of mRNA

Oxygen atom of furanose ring is a marked point of RNA reading for the ribosome complex: β(70S, 80S) = 4/5 = **O**/**H_2O**.

The movement of 40S ribosome along mRNA starts at the velocity β, but creates reading wave of the mark **CH_2** at the velocity *b*. As soon as 40S ribosome associates with 60S ribosome, the wave velocity of complex of ribosomes 80S is equal to the corpuscular velocity β. Wave velocity of tRNA is equal to b, but its corpuscular velocity is equal to the corpuscular velocity of ribosome β. Termination of translation begins when wave velocity of 80S ribosome is equal to b and exceeds its corpuscular velocity β: tRNA is pushed away from contour of protein synthesis by the ribosome.

Enzyme ρ-factor for prokaryotes has corpuscular and wave velocity $\rho_1(\beta)$, but moves with a delay with respect to 70S ribosome and is the marked point of terminator. As soon as the wave velocity of ρ-factor is equal to the wave velocity of tRNA, the processes of DNA transcription and RNA translation are terminated.

Function Ψ_U of tRNA

Ψ-Function of tRNA, which has four components Ψ_{37}, Ψ_{39}, Ψ_{43} and Ψ_{52}, was introduced in Chapter 1. Let us prove that Ψ-function of tRNA is a representation of CFAA metric.

Let us introduce the numeration of Ψ-function components with respect to the starting mark M(41) of RNA translation in CFAA metric $13^{+}7^{-}$:

$$\Psi_{37} = \Psi_{-4}\,(M); \Psi_{39} = \Psi_{-2}\,(M); \Psi_{43} = \Psi_{+2}\,(M); \Psi_{52} = \Psi_{+10}\,(M^*).$$

Then write the set of Ψ-function representatives on the basis of the genetic code triplets:

$$\begin{aligned}
\Psi_{-4}(M) &= \{F, I, I, K, L, N, Y, *\} \\
\Psi_{-2}(M) &= \{C, D, E, F, H, I, K, Y, L, L, L, M, N, Q, R, S, S, S, T, T, V, V, *, *\} \\
\Psi_{+2}(M) &= \{A, A, G, G, P, P, R, R\} \\
\Psi_{+10}(M^{-}*) &= \{A, A, C, D, E, G, G, H, L, L, P, P, Q, R, R, R, S, S, S, T, T, V, V, W\}
\end{aligned}$$

Function Ψ_{-4} (M) contains the third copy of the mark M:

$$\Psi_{-4}\,(41) = \mathbf{GF}83(K) = K(37).$$

Function Ψ_{-2} (M) contains the copy of the mark M.

Function Ψ_{+2} (M) contains the future mark M of the operator H(43):

$$Q_4\{+A + G - P^* + R\} = M(41).$$

$$Q_4[2\{+A + G + R\} - \mathbf{P} - \mathbf{P}] = \mathbf{GF}(83).$$

Amino acids A, G, P, R are the bundle over the base P, F, G, K.

Function $\Psi_{+10}(M^{-}*)$ is an evolutionary copy of the mark M, i.e., the starting mark M.

Let us select one representative of amino acid for each component of **the** Ψ-function taking into account the terminator* and designate their quantity as $Q_p(\Psi)$.

Then,

$$Q_p\{\Psi_{-4}(M)\} = 7; Q_p\{\Psi_{-2}(M)\} = 17; Q_p\{\Psi_{+2}(M)\} = 4^*; Q_p\{\Psi_{+10}(M^{-*})\} = 13.$$

From here

$$Q_p(\Psi) = Q_p(\Psi_{-4} + \Psi_{-2} + \Psi_{+2} + \Psi_{+10}{}^{-*}) = M(41).$$

Here we need to prove that the values of $Q_p(\Psi)$ correspond to the signature of signed and unsigned matrix 'α' of the conformal field of amino acids.

In the standard representation of tRNA, evolutionary sequence of mark M(41) looks like:

$$\Psi_{-4}(\mathbf{A}) \to \Psi_{-2}(\mathbf{U}) \to \Psi_{+2}(\mathbf{C}) \to \Psi_{+10}(\mathbf{G}^{-*})$$

and exactly corresponds to amino acid M(**AUG**). This is true because the triangle of differences of Ψ-function indexes induces 10-parametric Poincaré group only relative to marked point Ψ_{+2}(C).

Mark OH_4 of field ϕ

Mark $\mathbf{OH_4}$ splits on two marks **O** and $\mathbf{OH_2}$ in the metric 2^+2^- of field ϕ. Then the field ϕ forms a tetramer:

$$\phi_1{}^* = \{1\mathbf{C},{}^*\}; \phi_2{}^* = \{1\mathbf{O},{}^*\}; \phi_3 = \{1\mathbf{O}, 2\mathbf{H}\}; \phi_4{}^* = \{2\mathbf{C},{}^*\}.$$

Let us designate the quantity of differences in the row $Q_p\{Am\}$that are equal to $\delta = Q_p\{\mathbf{C}, \mathbf{O}, \mathbf{OH_2}, 2\mathbf{C}\}$ as $Q_\delta\{Am_1, Am_2\}$. We obtain CFAA metric 13^+7^-:

$$13^+ [\mathbf{OH_2}(Q_\delta = 4)\, \mathbf{O}(Q_\delta = 9)]\, 7^- [\mathbf{C}(Q_\delta = 5)\, 2\mathbf{C}(Q_\delta = 2)].$$

And even if the tumor supressor *p*53 controlls growth of field ϕ_3, it cannot stop the growth of field ϕ5(1**O**, 1**O**) with basic transition **T**66(I, L) → T25(C, T, V) mod M(41).

Literature Cited

1. Slavnov, A.A., Faddeev, L.L. 1970. Introduction to the quantum theory of calibration fields, Moscow, Nauka. p 240. (In Russian).
2. Atiyah, M., Hitchin, N. 1988. The Geometry and Dynamics of Magnetic Monopoles. Princeton University Press, Princeton, New Jersey.
3. Stcherbic, V.V. 2001. Calibration field of Hadamard matrices. Kiev: KMC "Poezija". p 116. (In Russian).
4. Stcherbic, V.V., Buchatsky, L.P. 2002. Mathematical structure of the parvoviruses protein coat. Visnyk problem biologii i medicini. No. 5. p 39-58. (In Russian).
5. Cramer, G., 1967. Leadbetter, M. Stationary and related stochastic processes. John Wiley.
6. Buchatsky, L.P. 1982. Molecular biology of parvoviruses., Kiev. KSU. p 83. (In Russian).
7. Afanasiev, B.N., Galyov, E.E., Buchatsky, L.P., Kozlov, Y.V. 1991. Nucleotide Sequence and Genomic Organization of Aedes Densonucleose Virus, Virology. 185, 323–336.
8. Mischenko, A.S. 1984. Vector bundles and their application. Moscow. Nauka. p 208. (In Russian).
9. Wess, J., Bagger, J. 1983. Supersymmetry and Supergravity. Princeton University Press, Princeton, New Jersey.
10. Lewin, B. 1987. Genes. John Wiley and Sons
11. Bashford, J.D., Tsohantjis, I., Jarvis, P.D. 1998. A supersymmetric model for the evolution of the genetic code. Proc. Nat. Acad. Sci. USA. Vol. 95. p. 987-992.
12. Hornos, J.E.M., Hornos, Y.M.M. 1993. Model for the evolution of the genetic code. Phys. Rev. Lett. Vol. 71. N26. p 4401-4404.
13. Bergman, P.G. 1942. Introduction to the theory of relativity. Prentice-Hall, inc, New York, p. 287.
14. Pauli, W. Jr. 1991. Theory of relativity. Moscow. Nauka. p 328. (In Russian).
15. Penrose, R., Rindler, W. 1986. Spinors and space-time. Spinor and twistor methods in space-time geometry. Vol. 2. Cambridge University Press, Cambridge.

16. Glagolev, N.A. 1963. Projecting geometry. Moscow. "Visshaja shkola". p 344. (In Russian).
17. Singer, M., Berg, P. 1991. Genes and genomes. University Science Books, Mill Valley, California.
18. Landau, L.D., Lifschitz, E.M. 1973. Field theory. Moscow. Nauka. p 504. (In Russian).
19. Rashevsky, P.K. 1936. Introduction to Riemann geometry and tensor analysis. Moscow. ONTI USSA. p 200. (In Russian).

Index